SECRETS OF THE SEA

BY CARL PROUJAN

Executive Coordinators: Beppie Harrison
                       John Mason
Design Director: Guenther Radtke
Editorial: Mary Senechal
           Gail Roberts
           Marjorie Dickens
Picture Editor: Peter Cook
Research: Patricia Quick
          Nadine Tobutt
Cartography by Geographical Projects

This edition specially produced in 1973
for International Learning Systems
Corporation Limited, London
by Aldus Books Limited, London.

Printed and bound in Yugoslavia by
Mladinska Knjiga, Ljubljana

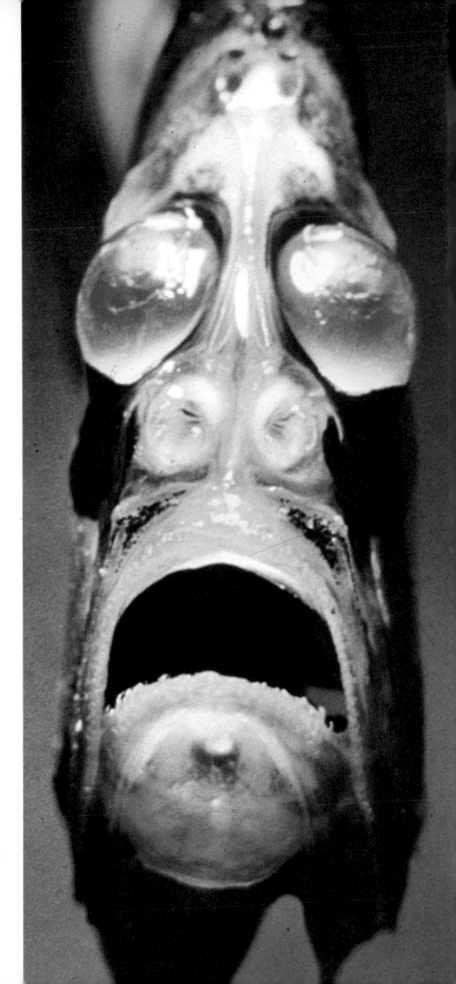

SECRETS OF THE SEA

DISCOVERY AND EXPLORATION

International Learning Systems Corporation Limited · London

Contents

Left: a deep-sea hatchet fish, *Argyropelecus hemigymnus,* one of the many grotesque animals that live in the darkness of the deep ocean. But this particular deep-sea monster is only $2\frac{1}{2}$ inches long.

Frontispiece: under the sea lies a mysterious world—a world whose landscape is even greater than that mirrored in its surface. For hundreds of years this world remained beyond the reach of man, but today the secrets it has hidden for so long are at last coming to light.

List of Maps

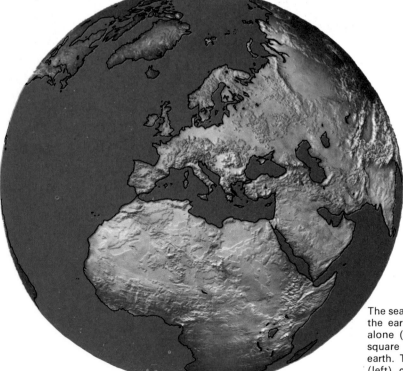

The sea covers more than 70 per cent of the earth's surface. The Pacific Ocean alone (right-hand globe) at 64 million square miles covers one third of the earth. The Atlantic and Indian oceans (left) cover nearly 60 million square miles between them.

The Last Great Frontier
1

At dawn on January 23, 1960, two men embarked on one of the most perilous journeys ever undertaken. Locked inside a steel sphere aboard the U.S. Navy's bathyscaph (deep-sea submersible) *Trieste,* Jacques Piccard and Donald Walsh set out to plunge through 35,800 feet of ocean in the deepest-known spot on earth. No one knew what they would find there, or whether they would come back alive.

The *Trieste* pitched and rolled on a stormy sea, watched by the anxious crews of two U.S. Navy escort ships. Far below them lay Challenger Deep in the Mariana Trench, a huge crescent-shaped gorge in the floor of the western Pacific Ocean. Inside their strange craft, in a space three feet across and less than six feet high, Piccard and Walsh waited, their eyes fixed on the depth gauge.

At 8:23 A.M. the signal came. The *Trieste* began to drop into the silence of the depths. It took nearly 40 minutes to reach 800 feet, but at 9 A.M. Piccard accelerated the rate of descent. First they passed through a twilight zone. Then darkness. Piccard flicked on the forward beam. As he peered into the sea, a flurry of tiny marine creatures streamed past.

Deep in the heart of the ocean, the two men were very much alone. A telephone provided their only link with the surface. This contact was reassuring. But Piccard and Walsh were far beyond the reach of assistance.

"9:20, depth 2,400 feet," reported Piccard. "Outside, total blackness. . . . We have entered the abyssal zone—the timeless world of eternal darkness." A chill penetrated the sphere as the temperature dropped rapidly. Thousands of tons of pressure from the surrounding sea gripped the descending craft. At 4,200 feet, the men were alarmed to see a thin trickle of water seeping in.

Black water rushed past as the *Trieste* shot on downward at 180 feet per minute. Piccard noticed with relief that the leak had stopped. At 20,000 feet the sphere began to plunge into the deepest ocean trench in the world.

At 29,150 feet Piccard noted "a vast emptiness beyond all comprehension." With perhaps a mile or more still to go, he was

Left: Donald Walsh (right) and Jacques Piccard in the bathyscaph *Trieste* after diving to the deepest-known part of the sea. They descended 35,800 feet, or nearly seven miles, into the Mariana Trench in the Pacific.

becoming increasingly worried about the moments ahead. At the bottom of the trench was a gap barely a mile in width. The *Trieste* would have to be right on target to avoid colliding with the rocky trench walls that might shatter it to pieces.

At 32,400 feet, there was a sudden explosion. The sphere shook as if it had been caught in a small earthquake. The two men looked anxiously at each other, fearing that they had hit the sea floor too fast and too soon. They waited and listened, but all was silent. After checking out the instruments, Piccard and Walsh agreed to continue the dive. They slowed their descent in preparation for touchdown on the ocean floor.

At 1:06 P.M., the *Trieste* landed gently on a bed of flat, snuff-colored ooze. Piccard and Walsh made "token claim, in the name of science and humanity, to the ultimate depth in all our oceans." But although the *Trieste* was resting on the bottom of the ocean, nearly

7 miles down, it was nearly 400 feet short of the ocean's deepest point at 36,198 feet.

Piccard and Walsh returned safely to the surface. Other daring explorers have sacrificed their lives in pursuit of the sea's secrets. What is it that lures men into the hostile environment of the underwater world? Why, in spite of all the difficulties and dangers, do men *want* to go down into the depths of the oceans?

For some divers, the difficulties and dangers are themselves reason enough. In an age when man has climbed to the peaks of the loftiest mountains, and crossed the frozen wastes of the poles, the depths of the sea remain unconquered. They form the last great frontier on earth.

Allied with the urge for adventure is a basic need to know. Men long to see for themselves what lies beneath the shimmering surface of the oceans. They want to know what it feels like to float in the watery world of *inner space,* the world of the ocean depths. Man hopes to solve the mystery of his own beginnings in the sea from which many scientists believe all life sprang.

But, today, curiosity and a desire for conquest are no longer the only reasons for probing the ocean depths. Man needs the sea if he is to continue to survive. If the population of the world keeps on growing as rapidly as it does now, the natural resources on dry land will run out. Under the sea lie vast unexploited supplies of food and minerals. The need to find and extract these resources has become the most urgent reason for man to explore the oceans.

The earth is a watery planet. More than two-thirds of it lie under the sea. The land masses are no more than islands in this immense watery mass. The oceans contain some 316 million cubic miles of water. If all the exposed land on the earth's surface were engulfed in the sea, the ocean floor would still be covered by an unbroken waste of water about two miles deep. The ocean bottom lies at an average depth of 12,450 feet. This is 4½ times as great as the average height of the land, including the mountains. At its deepest point, the

Above: a shoal of fish swimming over a coral formation. Man has always been fascinated by the life of the sea, but until recently, he had only been able to catch fleeting glimpses of this strange world.

Right: an artist's impression of the *Trieste* nearing the ocean floor. Protected in their steel sphere, the intrepid explorers look out onto the most remote landscape on earth.

ocean could accommodate the world's highest mountain with about $1\frac{1}{2}$ miles to spare.

But the sea is far more than a vast surface and a sheer mass of water. It is a complete world in itself. At shallower depths the scenery can be dramatic, but deeper down the waters are black and silent. The inhabitants of the underwater world are many and startlingly varied, both in appearance and size.

It was probably in search of living things that man first ventured into the sea. He began a relationship that has always contained

Right: the surface of the seabed is just as irregular as the surface of the land. Mountains rise from the floor of the ocean to appear as islands above the waves, and the sea-bed is traversed by great mountain ranges, or *ridges*. At places the ocean floor is cut by deep valleys called *trenches*. This map shows the principal ridges and trenches, and other features of the seabed.

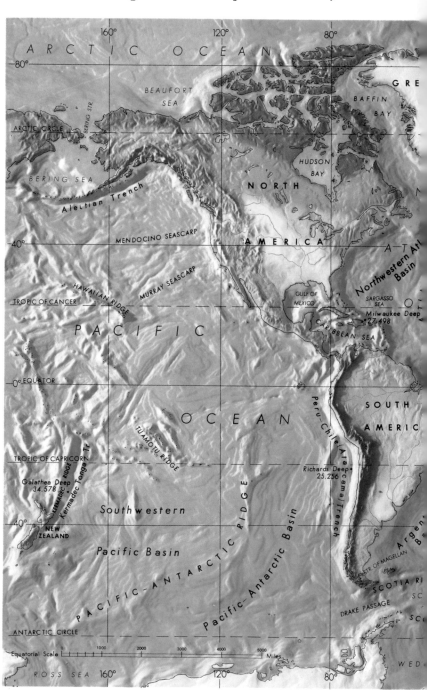

Above: an Akkadian seal (about 2300 B.C.) shows Zu the Birdman, being led before Ea, the Babylonian water god. The Babylonians, like other early civilizations, believed that gods controlled the forces of the sea.

elements of fear as well as wonder. Despite an early familiarity with the waters around the coast, the great depths of the open sea remained impenetrable and frightening. For centuries, the dark and shadowy realms of the deep were thought to be inhabited by terrible monsters. To early man it seemed that this must be where the immense forces of the sea lay hidden. And he ascribed these forces to the workings of powerful supernatural beings.

The first records of sea gods come from the Babylonian civilization, which flourished 5,000 years ago, though the myths about

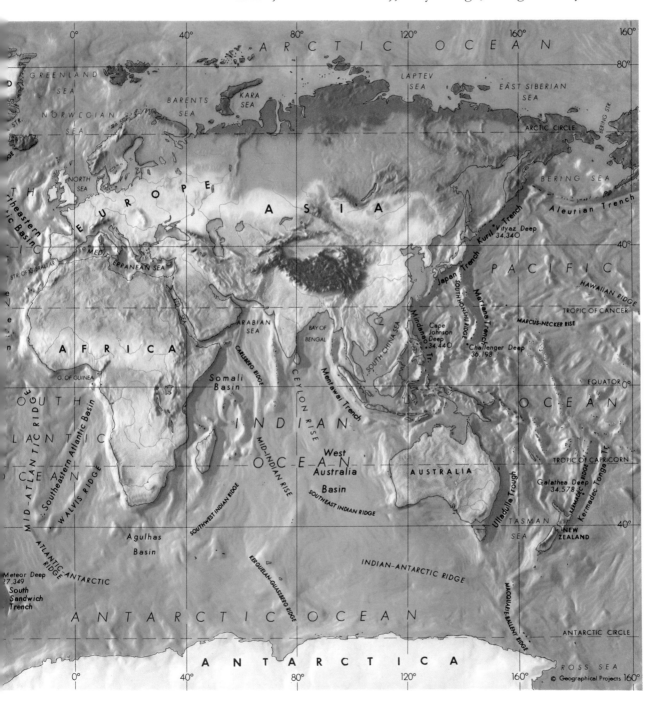

gods of the sea are even older. The Babylonians, who lived in what is now southeastern Iraq, worshiped a god named Ea, "the deity of the watery deep." Ea was a fresh-water god, but the Babylonians had sea gods and goddesses too, some of whom resembled the mermaids found in legends of the sea.

More plentiful evidence of the worship of underwater gods has been found on Minoan frescoes and drinking vessels. The Minoan civilization flourished from 3000 B.C. onward on the island of Crete in the Mediterranean Sea, a clear, warm, and largely tideless sea. Minoans sailed all over the Mediterranean to trade, and they were accomplished swimmers and divers. One of the most famous Minoan heroes was a diver called Glaucus who was said to have learned the secrets of the sea after eating a special kind of seaweed that enabled him to breathe underwater. The sea gods, impressed by his desire to visit the undersea world, made him immortal and he too became a god.

The Greeks, for whom Glaucus was also a god, inherited the Minoan tradition of diving. Among their gods were the most famous underwater deities of antiquity. The chief Greek god of the sea was Poseidon (later to be called Neptune by the Romans). In fits

Left: a French painting of the 1200's showing Alexander the Great (356–323 B.C.) at the bottom of the sea in a glass barrel. The story of his dive into the sea to observe marine life is probably only a legend which grew up long after his death. According to some versions of the story, he is reputed to have seen a monster which took three days to swim past his glass cage. Bibliotheque Royale Albert 1 er, Bruxelles. Ms. 11040, f.70v.

Right: an Indian painting of the 1500's illustrating the same legend. Notice how Alexander's appearance has changed from a typical European monarch with ermine and crown to that of a bearded Eastern potentate.

of anger, Poseidon was believed to beat the seas into fury with his trident. In an effort to placate him, the Greeks built temples in his honor at dangerous points along the coast.

The records and legends of ancient Greece contain many accounts of underwater exploits. Herodotus, the Greek historian, tells the story of Scyllias, an accomplished diver of the 500's B.C. Scyllias and his daughter Cyana sank at least one enemy ship by cutting its cables under the sea. They also recovered quantities of treasure from Persian ships that had been sunk by the Greeks.

Herodotus also mentions the use of submarine vessels in the 500's B.C. These were cages, made of glass, from which the approach of enemy ships could be sighted. Two centuries later, Alexander the Great is said to have observed marine life from a similar glass barrel, which was suspended into the sea by a golden chain. This story seems to have grown up long after Alexander's death and his submarine adventure probably never happened. But Alexander did use divers to saw through enemy defenses during the siege of the Phoenician city of Tyre in 332 B.C.

One of the most famous legends about warfare and the sea is the story of the "lost continent" of Atlantis, a large mythical island in the Atlantic Ocean. The armies of Atlantis were said to have planned to conquer the Mediterranean countries. They had made some conquests in Europe and Africa when they were defeated in battle by the Greeks. Later Atlantis was swallowed up in the depths of the sea during terrible earthquakes and floods.

Apart from their military exploits, the Greeks dived regularly for peaceful purposes, searching the sea for fish, coral, mother-of-pearl, and sponges. Greek sponge divers are known to have reached depths of 75 to 100 feet. These ventures into the sea were not limited by the air capacity of a diver's lungs. Aristotle, the Greek scientist and philosopher who lived in the 300's B.C., describes primitive diving bells used by sponge divers. The bells were weighted and filled with air. When out of breath, a diver would poke his head into the bell for air. Moreover, the air supply in the bell could be replenished with air delivered in weighted animal skins.

Above: a vase made in the 500's B.C. shows a Greek diver about to enter the sea. In their search for sponges, Greek divers are known to have reached depths of 75 to 100 feet, sometimes using primitive diving bells to replenish their air supplies. During their dives they observed other forms of marine life and so added to the growing knowledge about the sea.

These divers brought back knowledge as well as sponges. Their observations of underwater life were carefully noted by Aristotle. But Aristotle did not confine his studies to the reports of divers and fishermen. He sailed over large areas of the Aegean Sea in his quest for knowledge of marine life. He discovered, named, and described 116 kinds of fish, 24 kinds of crustaceans and marine worms, and 40 kinds of shellfish and *radiolarians*—minute creatures with delicate outer skeletons.

Being sailors, like most island peoples, the Minoans of Crete were the first to explore the sea. But their voyages were confined to the waters of their own familiar Mediterranean Sea. Rumor and superstition persisted about the world that lay beyond. However, by 600 B.C., the Phoenicians had sailed out into the Atlantic Ocean. They reached the British Isles and may even have sailed around Africa.

The first great individual explorer of the sea was Pytheas of Massalia. Pytheas, a Greek astronomer and mathematician as well as an explorer, sailed from Massalia (now the French port of Marseille) in about 325 B.C. Passing through the Strait of Gibraltar into the Atlantic, he made his way northward up the west coast of Portugal and past France and Britain. During a voyage which took him as far as the frozen seas of the Arctic, Pytheas made a number of scientific observations and was the first of the Greeks to suggest that the ebb and flow of the tides is related to the moon.

In the 100's B.C., Posidonius, a Greek philosopher born in Syria,

Left: a ship carved in a rock face at Lindos, on the island of Rhodes. The ship is Greek in design and the carving dates from the Hellenistic Age which began in 323 B.C. and lasted for nearly 200 years. During this period Greek culture spread into Egypt and throughout the Near East. In such ships the Greeks explored the Mediterranean and Atlantic.

set sail for Spain to confirm or refute the belief that, as the sun set in the west, it sank into the Atlantic and the sea sizzled with a hissing noise. Although Posidonius never heard the sun plunging into the ocean, he did note that the depth of the sea near the coast of Sardinia was 1,000 fathoms (6,000 feet). No one knows how this measurement was made, but it is a reasonable one. Depths in excess of 8,500 feet have been recorded near Sardinia by modern oceanographic vessels.

The Romans, like the Greeks, used divers in combat. But they appear to have had little other interest in what went on beneath the surface of the sea. Centuries passed, during which the art of diving was kept alive by the Arabs and by pearl divers of the East. Elsewhere, man withdrew from the challenge of the sea and turned his interest inland. The explorations begun by the Minoans and the Greeks ground to a halt and so did the science of oceanography.

Above: an Assyrian relief of about 750 B.C. showing men swimming underwater to attack a city. One of the attackers has an air-filled pig's bladder between his teeth. Some historians see this as an early attempt to provide air for an underwater swimmer. Others point out that the bladder would be too buoyant to keep the diver submerged, and think he is using it as a float.

Left: a portrait of Constantine John Phipps. Phipps made the first sounding of the deep ocean floor in 1773. Until then all attempts had failed, and many people still thought the sea was bottomless. Aboard H.M.S. *Racehorse* Phipps sounded the seabed between Iceland and Norway and recorded a depth of 683 fathoms. Below: a water color showing H.M.S. *Racehorse* and H.M.S. *Carcass* stuck in pack ice during the voyage they made toward the North Pole in 1773.

The Search Begins

2

The oceanographers of the ancient world had been puzzled by two questions in particular. How deep is the sea? How deep can animals live in it? Aristotle's works, lost and forgotten in the West during the Dark Ages from the 400's to the 900's, had been rediscovered by the 1300's. Although they helped to spark a new interest in the oceans' depths, it was not until the 1700's that a determined bid was made to answer the questions posed more than 2,000 years earlier.

The scientists who undertook this huge task had many problems to overcome. For hundreds of years, explorers had failed to sound (measure) the depth of the bottom in the open ocean. Their sounding lines were not long enough. Currents and surface winds prevented the lines from descending vertically, and on many occasions thousands of feet of line were paid out without touching the bottom. Many people continued to believe that the oceans were bottomless.

The first deep-sea sounding was made in 1773 by a British scientist named Constantine John Phipps aboard H.M.S. *Racehorse*. Using a weighted line, Phipps measured a depth of 683 fathoms (4,098 feet) between Iceland and Norway. This was to remain a record for 35 years.

While Phipps was making his soundings and measuring the temperature of deep water, other scientists were busy investigating underwater life. They captured their specimens with the aid of deep-sea dredges that were dragged across the bottom of the sea by a slow-moving ship. The first scientific dredging missions concentrated on relatively shallow water but, as the scientists scraped up living organisms from ever greater depths, undersea life became a controversial issue. Scientists began to take sides over the question of whether there was a depth beyond which life could not exist.

A French naturalist, François Péron, added to this controversy after returning from an around-the-world journey in 1804. During the trip, Péron had measured the temperature of the ocean at various depths. Because the water became colder and colder with greater depth, he was led to believe that the bed of the ocean was covered with "eternal ice," and that no life could possibly exist there. But there was little hard evidence to support or refute his contention.

Then, in 1818, a British sailor named John Ross (later Sir John), set sail for the Arctic Ocean on a voyage of discovery. Ross took with him a combination sounding and sampling device, which he called the *Deep-sea Clamm*. The Clamm could pick up a portion of

Above: an engraving of Sir John Ross being greeted by Eskimos of Prince Regent Inlet, at the northwest end of Baffin Island, in 1818. The ships in the bay are the *Isabella* and the *Alexander*. During this voyage Ross, commander of the *Isabella,* used the *Deep-sea Clamm* to retrieve a sample of the ocean bed from 6,000 feet. The sample contained living things, proving that life existed on the seabed.

the ocean's floor, and bring it back to the surface of the water.

On September 1, 1819, the Clamm was lowered over the side into the waters of the Atlantic. Fathom after fathom was counted off as the line played out into the sea. At 6,000 feet, the Clamm struck bottom and bit a six-pound chunk out of the ocean floor. Then Ross ordered the line to be hauled in. Throughout the long process, the men aboard the vessel wondered what the Clamm would hold.

Finally, the Clamm broke surface. From its teeth oozed soft mud. And in the mud there wriggled a tangle of tube-worms. Péron had been wrong. Here, at least, the floor of the ocean was not sheathed in ice. More important, it teemed with life. But was this the deepest part of the ocean? Were there depths still undiscovered where life could not exist? Many scientists believed so.

Among the most prominent of these scientists was a British naturalist named Edward Forbes. Forbes was born on the Isle of Man, a spot of land between England and Ireland, washed by the waters of the Irish Sea. As a boy, he had explored the cliffs and beaches that ringed his island home. Later, as a young student,

Above: a lithograph of the British naturalist Edward Forbes. In 1841 Forbes joined the British survey ship *Beacon* on a cruise to chart the Mediterranean and Aegean seas. During the voyage he dredged up animals from the sea floor 1,200 feet below him. What he found then, and in later studies, prompted him to classify marine life by dividing the sea into four principal zones.

Forbes often left the relative safety of the familiar coast to venture into the Irish Sea in a fishing boat.

Like Aristotle, Forbes observed, examined, and dissected the creatures he landed with dredge and trawl. Soon he had become one of the leading naturalists in Great Britain. He was an expert on the animals that dwelled in the waters of the Irish Sea. But the creatures that might lurk in deeper waters also intrigued him. So it was natural that he should jump at the opportunity to join the crew of the British survey ship *Beacon,* whose task it was to chart the waters of the Mediterranean and Aegean seas.

Forbes signed on as ship's naturalist in 1841. During the *Beacon*'s voyage, which lasted a year and a half, he dredged deeper than any scientist had done before him—1,380 feet down. From 1,200 feet, he brought up living starfish and shellfish that had previously been found only as fossils and were thought to be extinct.

From his findings on this journey and other studies, Forbes developed a system of classifying marine life. He divided the waters of the seas into four zones, each having a different *ecosystem* (animal

Left: mussels on a beach. These and other mollusks live between high- and low-water marks, on the edge of the sea. Forbes called this densely populated region the *littoral* zone, from the Latin word *littoralis*, which means seashore.

Above: a sea cucumber in eelgrass. This marine invertebrate is typical of the animals found in the sea to a depth of about 75 feet. This is Forbes' *laminarian* zone, named for the brown algae, *Laminaria*.

and plant populations and the environment in which they live).

The *littoral* zone consisted of shore waters between high-water and low-water marks. Living things in this zone were periodically bathed with air and sunlight. Seaweed flourished. The zone was alive with animals, especially mollusks.

The *laminarian* zone sloped to a depth of 60 to 75 feet below the low-water mark. This was the home of marine invertebrates (animals without backbones) who fed on vast "fields" of brown algae, and other plant food.

Probing below 75 feet, Forbes had found a layer of water crowded with large crustaceans and food fishes such as cod, haddock, and halibut. He named this the *coralline* zone. According to Forbes its lower boundary was marked by a depth of 300 feet.

Next, and seemingly bottomless, Forbes found the zone of deep-sea corals. And he was convinced that somewhere in this zone, pressure, darkness, and cold increased to a point where no life could possibly exist. Plants, which required sunlight, would vanish first, then animals that feed on plants would disappear. Finally, animals that prey on other animals would be lost. Forbes believed that this point would be reached at a depth of 1,800 feet. All the waters below that depth he called the *azoic* (lifeless) zone. But what of the worms found by John Ross at 6,000 feet? Forbes was either unaware of Ross's feat or he doubted that the worms were bottom dwellers.

Top right: a shark glides menacingly through the blue sea. The shark, together with many other fish and large crustaceans, lives in what Forbes called the *coralline* zone. He believed this extended from 75 to 300 feet, the lower limit of *coralline* algae growth.

Right: *Omosudis,* a deep-sea fish up to nine inches in length found below 3,000 feet. Forbes believed below 1,800 feet there was no life in the ocean. No plants or animals, he reasoned, could possibly survive the increasing coldness, darkness, and pressure. He called this region the *azoic* zone.

Right: Charles Wyville Thomson. As a former student of Edward Forbes, Thomson shared his belief that no life existed in the ocean below 1,800 feet. Between 1868 and 1870, however Thomson proved Forbes, himself, and many other scientists completely wrong. Even the slimy ooze he recovered from the bottom at depths of over 15,000 feet contained living things.

Below: Sir James Clark Ross. Between 1839 and 1843, Ross led an expedition to the Antarctic in H.M.S. *Erebus*. He was a scrupulous scientist and made many careful measurements of weather conditions, sea temperature, and the water depth. One sounding showed that the bottom depth was 14,500 feet, the greatest depth then recorded.

In 1860, 6 years after Forbes died at the age of 39, an event seemingly unrelated to the controversy over deep-sea life occurred. A telegraph cable between Corsica and Sardinia broke on the floor of the Mediterranean, 7,200 feet below the sea's surface. A repair crew was sent to the site of the break. After a considerable amount of work, the crew hauled the cable to the surface. To their amazement, a deep-sea coral clung to the cable at the point of the break. Corals are *sessile* animals—that is, they grow and develop while anchored to a solid object. They are not free swimmers. Therefore, the coral had fastened itself to the cable at a depth of 7,200 feet.

As foot by foot of the cable emerged from the sea, other sessile animals were found clinging to it. This was powerful evidence against Forbes' concept of a lifeless zone below 1,800 feet. Yet preconceived notions of such a lifeless zone lingered on. Many scientists could not conceive of living things spending their lives in total darkness, at temperatures close to the freezing point of water, and, more important, at pressures in excess of 1,000 pounds per square inch. (In the sea, water pressure increases at a rate of 0.442 pounds per square inch for each foot of depth.)

One of those who continued to believe in a lifeless zone was another British naturalist, Charles Wyville Thomson, who was later to make great contributions to the science of oceanography. Thomson, who had been a student of Forbes, decided to put his beliefs to

the test of controlled scientific observation. In 1868, he set sail on a small gunboat, H.M.S. *Lightning*, which had been loaned by the British Admiralty. Equipped with a deep-sea dredge, *Lightning* scraped the Atlantic floor from the Faeroe Islands, about 250 miles north of the Scottish coast, to as far south as Gibraltar. From depths as great as 3,600 feet, the dredge snared myriads of living creatures. In 1869 and 1870, Thomson cruised the Atlantic aboard another borrowed gunboat, H.M.S. *Porcupine*. From its decks, he lowered dredges deeper and deeper into the ocean. And always they came up loaded with living things—even from a depth of more than 15,000 feet. Life did exist in the total darkness of the deep. This part of the sea, where sunlight never penetrates, is known as the *abyss,* or *abyssal* zone. In most parts of the world, its upper boundary lies about 6,600 feet below the surface. Its lower boundary is the ocean floor. The abyssal zone is now known to be the world's largest ecosystem and covers half of the earth's area.

For Thomson, two questions remained unanswered. Although he had hauled living things from a depth of 15,000 feet, was there still a depth at which no living creature could survive? And if no depth was uninhabited, what sort of strange creatures would be found at depths yet unexplored? Thomson determined to dredge all the sea beds of the world, if necessary, until he found the answers, and he began to petition the British Admiralty for their support.

Above: a water color painted during Ross's voyage to Antarctica between 1839 and 1843 shows H.M.S. *Erebus* and her sister ship, H.M.S. *Terror* off the coast of Antarctica in 1841. Ross, commander of the *Erebus,* named the volcano he saw Mount Erebus.

While Forbes and Thomson were busy searching for life in the deep, plans were going ahead to lay an underwater telegraph cable between Europe and America. Such a project called for more efficient methods of measuring the ocean depths and increased knowledge of the physical properties of the sea and the nature of the ocean floor.

In 1839, Sir James Clark Ross, nephew of Sir John Ross, left for the frigid waters of the Antarctic aboard H.M.S. *Erebus*. Pausing on the way south, Ross ordered a boat lowered over the side. The boat carried a simple cargo—a 76-pound lead weight attached to a hemp line that was coiled around a huge reel. When the boat was a few yards from the mother ship, Ross commanded the oarsmen to drop the weight over the side.

The line spun rapidly from the reel, but it was nearly an hour before the weight struck bottom. It had carried 14,550 feet of line into the sea. Thus did James Clark Ross make the first abyssal sounding of "a depression of the bed of the ocean beneath its surface very little short of the elevation of Mont Blanc above it."

During his voyage, Ross also took many temperature readings at the ocean bottom. He found that the temperature at the bottom was the same at all latitudes—about 4°C. However, Ross had overlooked the effect of water pressure on the thermometer bulb. Pressure forced the mercury too far up the stem of the thermometer and gave a reading that was too high.

Wyville Thomson aboard the *Lightning* also measured the temperature of the sea and made a surprising discovery. He found that

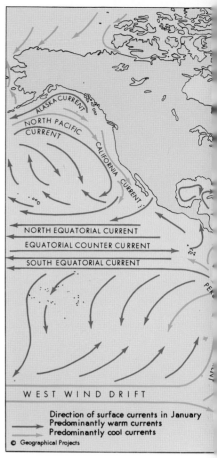

Above: the oceans of the world, showing the most important ocean currents. The currents—which can be either hot or cold—are great rivers in the oceans, set in motion by the winds. In the Northern Hemisphere the currents' flow is generally clockwise and in the Southern Hemisphere counterclockwise. This difference of direction is caused by the rotation of the earth on its axis.

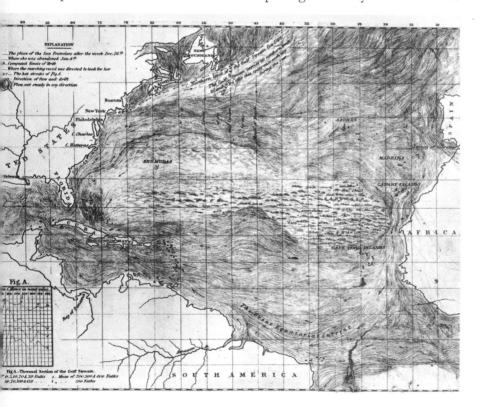

Left: a chart of the Gulf Stream from Matthew Maury's book *Physical Geography of the Sea and its Meteorology,* published in 1855. Maury believed that warm winds from the eastern branch of the Gulf Stream (the North Atlantic Current) caused the relatively mild climate of Britain and western Europe.

Right: Matthew Fontaine Maury. His *Wind and Current Charts* formed the basis of the government's pilot charts.

"great masses of water at different temperatures (and depths) are moving about, each in its particular course; maintaining a remarkable system of oceanic circulation, and yet keeping so distinct from one another that an hour's sail may be sufficient to pass from the extreme of heat to the extreme of cold."

Others had charted such "rivers" in the sea. Nantucket whaling captain Timothy Folger charted the Gulf Stream in 1770. And in the same year the great American scientist and statesman Benjamin Franklin drew up temperature tables of the Gulf Stream so that a ship's navigator could tell whether he was in or out of its northeast-ward-flowing current. In 1781, Charles Blagden charted the cold Labrador Current, which flows from the Arctic Ocean and runs into the warm Gulf Stream near the Grand Banks off Newfoundland. Where the warm, moisture-laden air of the Gulf Stream meets the cold Labrador Current, the sea is often shrouded in dense fog.

The work of Folger, Franklin, Blagden and others was expanded into a world-wide system of meteorological observation by Matthew Fontaine Maury, who has been called the *Pathfinder of the Seas*. As a young officer in the U.S. Navy, Maury began to study winds and currents as a means of shortening the journeys of ships across vast expanses of the sea. He took lengthy notes wherever he sailed, and drew graphs and charts of all he observed. Then, in 1839, after 14

Above: the deep-sea sounding device developed by Maury and Brooke. It consisted of a hollow tube running through a cannon ball. When lowered, the ball was held in place by a collar fixed to two hinged brackets.
Below: when the tube struck bottom the drop in tension on the sounding line tripped the brackets and the ball was released. The tube, with a sample of the seabed, was then reeled in.

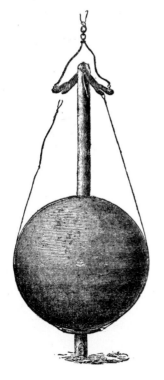

years of sea duty, Maury was lamed in an accident and declared unfit for active service.

But the Navy did not waste Maury's talents. In 1842, he was put in charge of the Depot of Charts and Instruments. Now Maury pored over the records of countless voyages in his search for "short cuts" across the seas. With these in hand, Maury charted the tracks of many ships making identical voyages. He noted the season in which a voyage was made. And he recorded wind and current data reported from each voyage. He even asked that navigators indicate temperatures, barometric pressures, encounters with fog, the sighting of whales, birds, and islands.

During the next 10 years, Maury accumulated data obtained from 265,298 days of observation—the equivalent of almost 727 years of sailing by a single seafarer. From this information, he made charts indicating the prevailing winds and currents at different times of the year, which could be used to find the speediest routes across the seas.

But Maury was not yet satisfied with the feat he had accomplished. In 1853, he sailed to Belgium to attend a conference where he would meet representatives of all the world's maritime nations. Maury knew that to make ocean voyages safe as well as speedy, a world-wide system of meteorological observations would have to be set up. Addressing the conference, he pleaded for the cooperation of scientists throughout the world. The representatives of the seagoing nations agreed to support Maury's scheme. The foundation of today's weather bureaus had been laid.

Maury was also a physical oceanographer—he studied the basic processes at work in the sea. In his famous book, *Physical Geography of the Sea and Its Meteorology,* published in 1855, Maury presented many fascinating concepts and theories. For example, he suggested that ocean currents influence climate. He believed that warm winds from the eastward branch of the Gulf Stream, called the North Atlantic Current, raise the temperature of Britain and western Europe by 20°F over that of other land areas at similar latitudes. To support this contention, Maury noted that Britain, bathed by the North Atlantic Current, is a land of greenery in winter, when Newfoundland, at much the same latitude but gripped by the cold Labrador Current, is covered in ice and snow. He also suggested that currents such as the Gulf Stream spawn destructive storms. And he stated that the fogs of Newfoundland were caused by the condensation of moisture above the Gulf Stream. His theory was that the moisture condensed when the cold temperatures of the southward-sweeping Labrador Current cooled the damp air of the Gulf Stream.

Maury also tried to explain the forces that cause currents to flow. He concluded that many factors played a role. Among these are winds, differences in the density of water, and the earth's rotation.

Maury was involved in practical as well as theoretical oceanography. Together with John M. Brooke, a young naval officer, he developed a new sounding device that could measure depth and bring back a small sample of the seabed. The device was mostly

Brooke's invention but it was Maury who arranged for ships all over the world to use it. By organizing a series of systematic soundings of the North Atlantic, Maury played a leading part in the laying of the first transatlantic telegraph cable.

The successful laying of this cable on July 27, 1866, gave rise to a new wave of interest in oceanography. Six years later, the Royal Society of London charged Wyville Thomson with the task of learning "everything about the sea." This was the chance that Thomson had been waiting for, and he was determined not to return empty-handed.

Above: the steamship *Great Eastern*. This ship laid the first successful transatlantic telegraph cable in 1866. It was the only ship of its time capable of carrying the 2,300 nautical miles of cable needed to traverse the Atlantic. The laying of the telegraph cable had been made possible by Matthew Maury's soundings of the Atlantic Ocean floor. Below: crewmen of the *Great Eastern* splicing the cable after a fault developed in July, 1865. The cable broke and the attempt was abandoned.

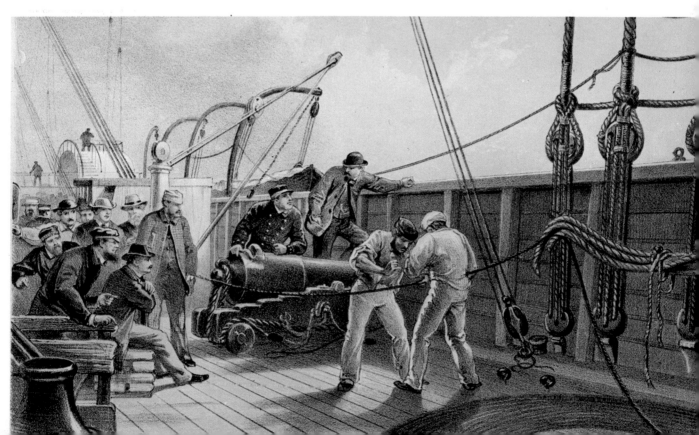

Challenger Faces the Unknown

3

Left: water color of H.M.S. *Challenger.* In December, 1872, this 2,306-ton corvette began a world-wide voyage of scientific research and discovery. The team of scientists on board was headed by Charles Wyville Thomson. The information they collected during the 3½-year voyage laid the foundation of the era of modern oceanography.

On December 7, 1872, a British man-of-war edged from its berth at Sheerness, about 30 miles east of London. It was making for Portsmouth, on the south coast of England, where it would start man's first global oceanographic expedition—an expedition that still holds the record for the longest continuous scientific mission. To the idle passer-by, the ship must have seemed a strange sight. For only 2 of its 18 cannons poked through its gun ports. The rest had been removed to make room for scientific equipment. This ship was not outfitted to do battle with a human enemy. Its only foes during a 3½-year voyage would be winter and rough weather. It would log 68,890 nautical miles in an adventure that would take it through all the world's oceans, except the Arctic. Its commander was George S. Nares of the Royal Navy. But, in a very real sense, the man at the helm was Charles Wyville Thomson.

The vessel was the 2,306-ton corvette H.M.S. *Challenger.* Though it would make most of its journey under sail, it was equipped with an auxiliary steam engine. And it was aptly named. It was on loan from the British Admiralty to the Royal Society of London, which had pleaded for a ship to challenge the unknown reaches of the oceans' depths.

Thomson headed the small scientific team aboard the *Challenger,* which included four naturalists, John Murray, H. N. Moseley,

Below: scientists using microscopes to examine biological samples during research on board H.M.S. *Challenger.*

Right: the dredging and sounding equipment on board H.M.S. *Challenger*. Sounding weights, dredges, and trawls were let down into the sea over the side of the ship. They were reeled in on drums powered by two steam engines near the mainmast. A system of blocks and spring units reduced the strain on the drums during rewinding.

Above: John Murray, one of the four naturalists on the *Challenger* expedition. He was particularly interested in the formation and composition of deep-sea sediments. Through studying material brought up in the *Challenger's* dredges and sample tubes, he began to distinguish two main types of sediment. The classification he devised for these sediments is still used today.

Rudolph von Willemoes-Suhm, and J. J. Wild, and one chemist, J. Y. Buchanan. During the voyage, the researchers would work in two well-equipped laboratories—one designed for analyzing samples of seawater, the other built for the study of animals and plants.

Storerooms below were crammed with bottles to take samples of water from various depths. There were miles of sounding line that would be used to measure the depths of the world's oceans. Dredges and trawls lay ready to scrape the floor of four oceans, snaring animals that had never been seen before. Some of the scientific gear was primitive by modern standards. For example, the sounding lines were made of hemp, a plant fiber. The end of each line was tied securely to a 200-pound lead ball. And the line itself was wound around a drum 10 feet in diameter. When a sounding was to be made, the *Challenger* was put under steam power so that it could be held fairly stationary, against the pull of winds and ocean currents. The weight was dropped over the side, unreeling the line from the drum.

Along the line, marks had been made every 600 feet. As the line was carried into the sea, one of the scientists noted the number of marks that slipped below the surface. However, during early soundings, an unforeseen problem arose. The line would continue to unreel from the drum even after the lead ball had struck bottom.

Apparently, the weight of the line that had uncoiled into the ocean was sufficient to drag more line from the drum. To solve this problem, which threatened to affect the accuracy of deep-sea soundings, the *Challenger* scientists took advantage of a principle developed in the 1600's by the English scientist Robert Hooke.

Hooke had invented a device for measuring depth without a line. The device consisted of a wooden ball connected to a lead ball. The two balls were thrown over the side of a stationary vessel. When the lead ball hit bottom, the wooden one was automatically released by a latch that opened on impact. The wooden ball then rose to the surface.

By trial and error, Hooke discovered how long the ball took to hit the bottom. From this he found that the time that elapsed between the dropping of the device into the water and the reappearance of the wooden ball could be used to measure depth. Knowing that there was a predictable relationship between rate of descent and depth, the *Challenger* scientists timed the rate at which their line unwound into the sea. When the rate began to slow down, the scientists assumed the lead ball had struck bottom. At great depths such soundings could take several hours. Modern sonar techniques, in which a *pulse* of sound is bounced off the bottom, can make them in seconds.

Using line and ball sounding methods, the *Challenger* plumbed a depth of 26,850 feet near the Mariana Islands in the western Pacific. As far as we know, the sea floor in this area of the Pacific is deeper than that of any other ocean bed in the world. A record depth of 36,198 has been measured in Challenger Deep, which is in the

Above: the dredge used by *Challenger*. It consisted of a twine-netting bag held open by a rectangular iron frame. Below: once the dredge was trailing behind the ship the crew sent a "traveler" weight down the line. It stopped at a toggle tied to the rope and carried the dredge to the bottom.

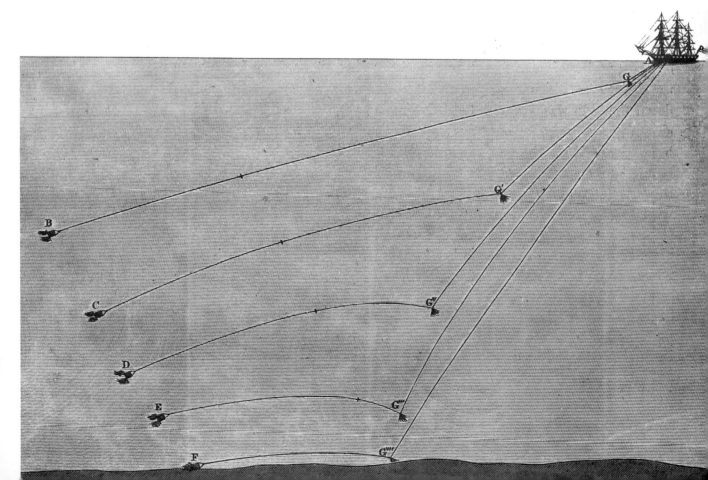

Mariana Trench, 200 miles southwest of Guam in the Mariana Islands.

What of life at great depths? The *Challenger*'s trawls and dredges brought up thousands of animals and plants. Of the animals, 4,717 species and 715 genera (groups of species) had never before been seen by man. Moreover, at no depth were there no living creatures. And the *Challenger* dredged the bottom at depths greater than 19,000 feet. But though animal life seemed to know no bounds, plants were not found beneath about 600 feet.

The *Challenger* scientists also found some explanation of how living things can survive in the cold, dark, high-pressure environment of the deep.

One creature, dredged from below 12,000 feet, was found to be equipped with its own light source—a powerful light-emitting organ. Some had huge eyes that could undoubtedly sense the faintest flicker of light. Still others had no eyes at all. They relied on sensitive organs of touch such as delicate antennae.

How did such animals survive the pressure of more than 8,400 pounds per square inch that exists at a depth of about 19,000 feet? An experiment performed by the expedition's chemist, J. Y. Buchanan, provided the answer.

Buchanan sealed both ends of a length of glass tubing, which he ordinarily used for experiments aboard ship. He wrapped the air-filled vial in cloth and placed it in a perforated copper container. Then the container and its contents were lowered over the side to a depth of 12,000 feet. Finally, the container was hauled back up and examined.

Buchanan found that the copper shell had been crushed. And of the glass tube it had held, nothing remained but powder. The perforated copper container was normally used to house thermometers that had measured water temperature at great depths. From past experience, Buchanan knew that the container could survive undamaged from depths greater than 12,000 feet. Yet it had not survived this plunge.

Buchanan had a reasonable, and correct, explanation. The glass tube, sealed at the sea's surface, contained air at atmospheric pressure—14.7 pounds per square inch. At some depth, the pressure outside of the tube so much exceeded that inside, that the tube collapsed. For an instant, there was a hollow space within the container. Before water could rush through the perforations in the container, the enormous water pressure drove the walls of the container inward.

Normally, when such a container was lowered into the sea, water passed through the perforations and equalized the pressure on both sides of the container's wall. The thermometer inside the container, was essentially a solid object. Therefore, it did not collapse. Water pressure, though enormous, remained balanced inside and outside of the container and no damage resulted.

This experiment, reasoned Buchanan and his colleagues, explained how animals survived under great pressure. The fluids inside their

Above: two crewmen empty the dredge net on board the *Challenger.* The closely woven netting around the bottom of the dredge held everything but the finest mud, so the scientists could study even very tiny animals.

Above: the deep-sea hatchet fish (*Argyropelecus sp.* three inches long), one of the many light-producing marine animals the *Challenger* scientists encountered. Left: the light organs on the underside of the fish. Light is produced in them by a complex chemical reaction. In the darkness of the deep ocean these pinpoints of light serve as recognition patterns and may also attract prey. Below: the well-equipped natural history workroom on H.M.S. *Challenger*.

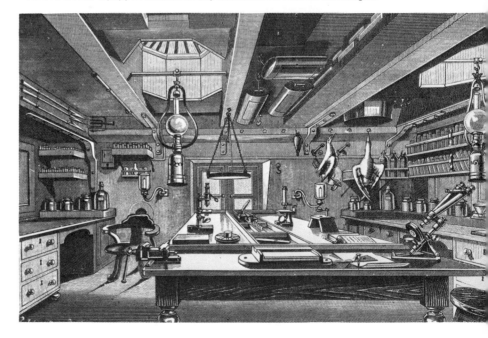

bodies were at the same pressure as was the water outside. Armed with this concept, the *Challenger* scientists concluded that life would be found at any depth, no matter how great. The one remaining factor —cold temperature—would affect only the kind of life present, not the existence of life itself.

Challenger scientists discovered the greatest number of marine living things in the relatively cold waters of the South Atlantic. These animals thrived in a food pyramid whose base consisted of an immense surface carpet of floating microscopic plants and animals that have since come to be known as *plankton* (from the Greek, meaning *wandering*). Plant plankton, mostly diatoms (single-celled algae), utilize sunlight to manufacture food from carbon dioxide, minerals, and water. These plants are the food of animal plankton, which, in turn, are the food of higher animals, and so on up the pyramid to large fish and aquatic carnivorous mammals.

Although questions concerning life in the deep were both intriguing and exciting, the 3½-year voyage of the *Challenger* also shed light on the riddles posed by physical oceanographers. Temperature readings were made at the 362 *stations* (stops) made by the *Challenger*. Self-closing bottles were lowered to various depths at each station to get samples of the water. A sounding was also made on each occasion, and a sample of the bottom was taken. From all the data, a picture of the sea's depths began to develop. In all, *Challenger* charted 140 million square miles of ocean floor.

Bottom temperatures of the abyss were almost always found to be near the freezing point of water. The temperatures were measured by a *self-registering thermometer* developed by Wyville Thomson during the *Lightning* expedition. The instrument recorded only the maximum and minimum temperatures to which it was exposed. A number of these thermometers, each housed in a protective perforated copper case, were attached to a sounding line at various intervals. The line was lowered into the sea to a known depth and then hauled up. Readings were taken of each thermometer and plotted on a graph. This graph provided a temperature *profile* at any given location. Similar instruments were lowered at each station to determine water pressure and water density at various depths.

Another discovery made by the *Challenger* team, which has great significance today, was that, in many places, the ocean floor is littered with manganese nodules. The nodules vary in size from

Above: crew members on *Challenger* retrieving a thermometer from the sea. Below: the thermometer they used. It registered only maximum and minimum temperatures and was housed in a protective perforated copper case (right).

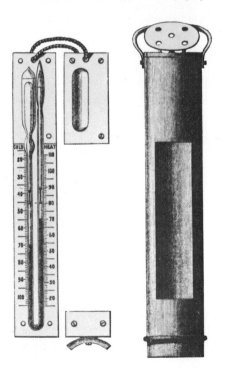

Above: a simplified food pyramid. The abundance of life that the *Challenger* scientists discovered in the sea all depends for its food supply on the tiny plant *plankton* (from the Greek, meaning *wandering*) that floats near the surface. Plant plankton produces food by photosynthesis. It is eaten by animal plankton, which is in turn eaten by higher animals. The whale shown above feeds on the animal plankton and itself provides food for man. The plants and animals forming plankton are so small that most of them can only be studied under a microscope.

Left: a sample of animal plankton.

Right: a sample of plant plankton.

about 0.5 centimeters to 25 centimeters. In addition to manganese, they are rich in iron and contain significant amounts of aluminum and magnesium, as well as nickel, copper, and cobalt. Some of these nodule beds are at present being tapped by oceanographic mining firms.

While studying bottom samples John Murray developed a system for classifying sediments in the abyss that is still used today. Murray studied the bottom samples brought up by *Challenger* devices, and divided the sediments into two broad categories—*pelagic* sediments

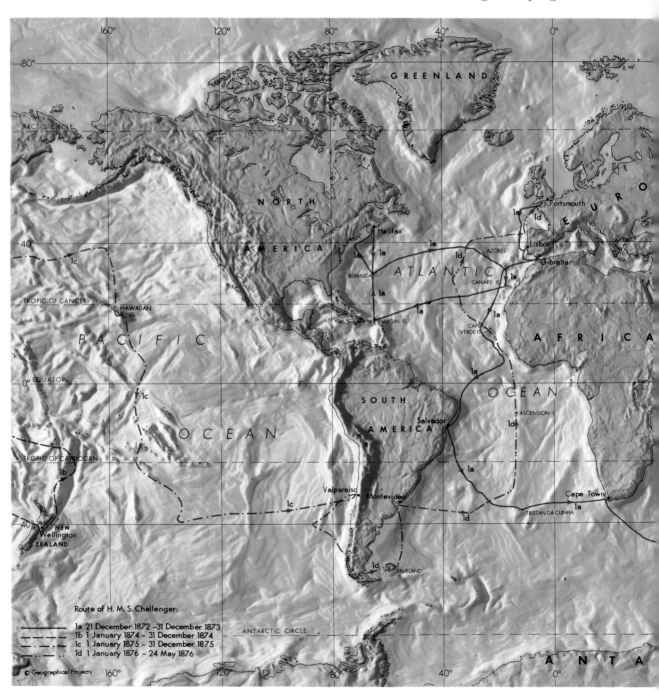

Route of H. M. S. Challenger:

———————— 1a 21 December 1872 –31 December 1873
– – – – – – 1b 1 January 1874 – 31 December 1874
– · – · – · – 1c 1 January 1875 – 31 December 1875
– ·· – ·· – ·· 1d 1 January 1876 – 24 May 1876

© Geographical Projects

and *terrigenous* sediments. According to Murray's classification, the pelagic sediment consisted of fine-grained material that rides waves, winds, and currents of the sea, and falls as a constant rain on the ocean floor. This material is made up of inorganic red clay or has its origin in organic (once living) material. The terrigenous sediment consists of particles of various sizes that come in the main from nearby continents or islands. Murray subdivided this later category into: blue, green, and red muds; volcanic mud; and coral sand and mud.

Left: the voyage of H.M.S. *Challenger* between 1872 and 1876 was the first global oceanographic expedition. Commissioned by the Royal Society of London to find out "everything about the sea," the expedition circled the earth, visiting every ocean except the Arctic. The data brought home by the expedition was to fill 50 volumes of the *Challenger Reports*.

Analysis of seawater brought up from various depths at the 362 stations revealed that the waters of the world's oceans have a constant composition of dissolved substances. That is, the proportion of substances in any sample of seawater will be approximately the same regardless of the origin of the sample. But the salinity (concentration of salt) varies at different locations and depths. In general, the parts of the ocean that receive heavy rainfall and those near the mouths of great rivers are less salty than average. In regions where the sun and wind cause a high rate of evaporation of moisture from the surface, the salinity is usually high.

The topography (surface features) of the cauldrons that hold the vast salty seas was also studied on the *Challenger* expedition. Thomson and his colleagues were familiar with the often precipitous contours of the earth's continents—mountains rising sharply from deserts and plateaus, and sheer cliffs plunging into the sea. But beneath the oceans the topography appeared to be different. Over thousands of years, a gentle rain of sediments had softened the contours of the ocean floor.

In May, 1876, the *Challenger* made its way back along the English Channel toward Portsmouth. The epic voyage was drawing to a close. It had cost the lives of two men. Rudolph von Willemoes-Suhm had died of *erysipelas* (a skin disease), and a young sailor had been swept overboard, caught in the line of a descending dredge. But much had been gained. The data brought home by the *Challenger* filled 50 volumes containing 29,500 pages. These reports were to launch the era of modern oceanography and came to be called "the oceanographer's Bible."

Their impact on the scientific world was immediate. Countries all over the globe began to fit out expeditions to follow in the wake of the *Challenger*. Other pioneer oceanographers were soon making their own contributions to the young science.

Among them was Alexander Agassiz, a Swiss-American scientist who had been a member of the *Challenger* team and had written

Left: a scientist emptying a water sample bottle on board *Challenger*. The device was lowered into the sea with its sample compartment open. At the desired depth, or when the device hit the bottom, a sudden jerk on the rope released a cylinder which slid down, sealing off compartment and sample.

two volumes of the *Challenger Reports*. He spent 25 years, from 1877 to 1902, in further exploration of the undersea world. Sailing aboard the *Blake* and the *Albatross,* he made extensive soundings in the Caribbean, the Indian Ocean, and the tropical Pacific. Using improved dredges and trawls he snared creatures from depths between 600 and 14,000 feet. He found that the most populated underwater zone was that between 1,200 and 12,000 feet. And within this zone, there was a 4,200-foot layer (1,800 to 6,000 feet) that harbored the greatest variety of animal species.

Another man who dedicated much of his life to the study of the seas was Prince Albert I of Monaco. Between 1885 and 1915, Prince Albert sailed on expeditions in the Mediterranean and Atlantic aboard his own yachts. He wrote many books about his work and helped to bring oceanography to the attention of the general public. In 1910, he founded one of the world's most famous marine museums at Monaco.

Below: a drawing made on board of a young fish (*Antennarius sp.* which is three inches long) found entangled in weed in mid-Atlantic on March 6, 1873.

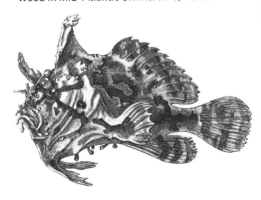

43

Above: the Norwegian explorer Fridtjof Nansen at the bow of his ship, the *Fram,* in the summer of 1894. The *Fram* was specially built to withstand the polar ice floes on this expedition to explore the North Polar basin. When, as Nansen had expected, it became locked in the ice, the men on board were able to make scientific studies of polar conditions.

Although these early pioneers had restricted most of their efforts to relatively warm seas, the vast regions of frigid Arctic water did not go unexplored. In 1893, the 128-foot-long ship *Fram* weighed anchor at Pepperviken, Norway. Under the command of Fridtjof Nansen, a Norwegian explorer with a degree in zoology, the *Fram* sailed northward.

For three years, Nansen struggled against the hostile environment of the Arctic—first aboard the *Fram,* and later on foot trekking across the endless ice. Early in the voyage, the *Fram* became locked in the ice. There she remained for 35 months, while her crew performed various scientific tasks. Surface and water temperatures were taken. Soundings were made through holes cut in the ice. Meteorological conditions were studied. And the position of the ship, held fast in the drifting ice pack, was determined every second day.

Nansen discovered that, contrary to popular belief, the north Arctic waters are not shallow. Some of his soundings struck bottom at depths greater than 12,000 feet. Nor were these waters devoid of life. In May, 1894, the *Fram* crew observed for the first time what has come to be called *the early summer plankton bloom.* Beneath the melting ice, a population explosion was occurring.

Uncountable swarms of diatoms were reproducing and forming brown patches on the ice. Algae began to appear as did microscopic animals. The base of a food pyramid was being forged, one that could support whales, fish, and other creatures of the Arctic.

On March 14, 1895, Nansen set out with a companion in an attempt to sledge his way to the North Pole. At the time, the *Fram* was at 83° 47′ north—about 483 miles from the pole. The two explorers came within 272 miles of the most northern spot on earth—no one had come closer—before snow and ice forced them to turn back.

In August, 1895, they reached Franz Josef Land. There they remained for nine bitterly cold months of frozen solitude before being rescued by a British polar expedition. Meanwhile, the *Fram* had broken loose from the grip of the polar ice. On August 20, 1896, she sailed into the port of Skjaervo, Norway, where Nansen was waiting to meet her.

Nansen's exploits showed that oceanographers were ready to challenge the greatest hazards on the surface of the sea in the search for knowledge of its depths. But the greatest test of man's courage and ingenuity remained—would he descend deep into the black abyss to see firsthand what wonders it held?

Above: a pastel drawing by Nansen showing the midnight sun near the North Pole. In the extreme north in summer, it is never dark and the sun shines through the night.

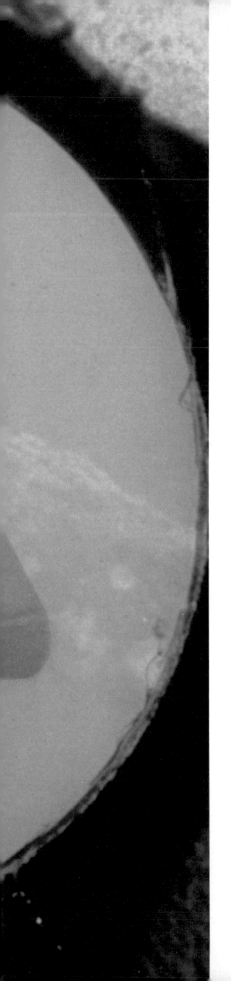

Under the Sea
4

Man is out of his element in water. He cannot swim around under the sea for more than two or three minutes without returning to the surface for a fresh gulp of air. The Greek divers of classic times achieved record dives of $4\frac{1}{2}$ minutes without breathing devices, but only in relatively shallow water and on condition that they did not move at all. When, in the 1300's and 1400's, men turned their attention again to the sea, their first interest lay in providing the diver with the apparatus necessary to breathe and move with ease at great depths and for considerable periods of time.

Among the first thinkers to tackle this problem was Leonardo da Vinci, one of the most brilliant men of the Renaissance (the period of artistic and scientific inquiry in Europe which lasted from the 1300's to the 1500's). His designs for underwater equipment included a headpiece of rigid leather to resist the pressure of water. Fitted to it was a breathing tube topped by a cork float to keep it above the surface. Glass lenses covered the eyeholes and the helmet was equipped with a series of spikes to ward off underwater monsters.

Leonardo also designed webbed gloves and flippers, but he never tried out the equipment he invented. If he had, he would have discovered an unexpected drawback. Equipped with Leonardo's breathing tube, a diver would have suffocated at a depth of little

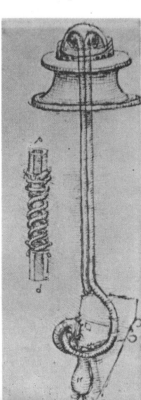

Left: a diver photographed through the porthole of the U.S. Navy's underwater habitat *Sealab II*. At a depth of 206 feet, he is completely cut off from the surface. However, the sophisticated equipment he is wearing enables him to move freely and safely in the environment of the ocean.

Right: a sketch of a device for breathing underwater made by the artist Leonardo da Vinci (1452–1519). The design shows a mask fitted with breathing tubes leading to a surface float.

more than five feet. The pressure of water on his chest at such a depth would prevent his lungs from expanding sufficiently to inhale the air in the tube.

Not until 1690 did Edmund Halley, who discovered the famous comet that bears his name, develop a means of piping pressurized air to a diver in a bell. If the air the diver breathes is at the same pressure as the water surrounding him, his chest is not crushed and he can breathe normally. This is the principle on which the diving bell is based. A diving bell is a container the bottom of which is open to the sea and which is large enough to hold one or two divers. Once the bell is submerged, the weight of the water compresses the air inside the bell and the men are able to breathe for some time. However, the farther the bell descends, the more water will rise in it and the less air will be available. The increasing water pressure

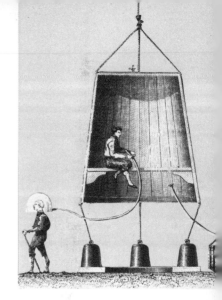

Left: The diving bell invented in 1690 by Edmund Halley. As the volume of air trapped inside decreased with increasing depth, extra air was piped in from casks suspended outside the bell at pressures equal to that of the water surrounding the bell. One of the divers shown here is using an individual bell supplied with air from the main bell. This small bell was the first practicable diving suit.

Below: a more sophisticated diving bell of the mid-1800's. Air was supplied to the men inside the bell from a hand-driven pump at the surface.

compresses the air within the bell, and in so doing raises the air pressure. For example, at a depth of 33 feet, the air within the bell is compressed to half its original volume and the pressure is doubled from 1 atmosphere (14.7 pounds per square inch), to 2 atmospheres (29.4 pounds per square inch). At 66 feet, air volume is reduced to one-third of what it was originally and the pressure is tripled to 3 atmospheres. For every additional 33 feet in depth, the pressure increases by 1 atmosphere. The volume of air in the bell decreases also, but not uniformly. At the relatively shallow depth of 627 feet, the volume of air in the bell would be compressed to one-twentieth of what it had been originally. And the pressure would be 20 atmospheres (294 pounds per square inch).

Assuming that the bell were 10 feet tall, the open space within it at a depth of 627 feet would be only 6 inches deep—hardly enough room to work in, let alone keep instruments functioning. There is a solution to this space problem. By pumping air into the bell from the surface at pressures equal to that of the water surrounding the bell, water can be kept from rising inside it. This was the technique discovered by Halley. He attached two empty casks to a diving bell by means of flexible tubes. A hole in the bottom of each cask let in water, whose pressure drove air from the cask, through the tube and into the bell. The divers inside the bell could let in this air when needed, simply by turning a tap.

Halley himself tested his invention on several occasions and stayed "at the bottom, in 9 or 10 fathoms [54–60 feet] of water, for above an hour and a half at a time, without any sort of ill consequence . . ."

Halley did find one inconvenience in his bell, which he was quick to remedy. Inside the bell was a bench on which the divers remained seated. This meant that they had a very narrow area of vision. Halley's solution was to provide the divers with small, individual bells, attached to their heads and linked by breathing tubes to the main bell. In this way, he invented the first workable diving suit.

Once it was realized that a man could not breathe the ordinary air of the atmosphere when his lungs were under pressure from water, protection of the diver from this hazard of pressure became the prime consideration of diving-suit designers. They began to devise unwieldly, armor-like suits, often reinforced with metal.

One of the first men to use this type of diving suit was the Englishman John Lethbridge, who had a long career as a diver for treasure from sunken wrecks. Lethbridge's diving apparatus consisted of a barrel-like suit, bound with iron hoops and fitted with leather sleeves. Once Lethbridge was inside the barrel it was bolted behind him, and his only means of vision was a four-inch diameter porthole. He was lowered into the sea by a cable and the air in the barrel enabled him to stay down for about half an hour. His assistants would then haul him to the surface and pump fresh air to him from a pair of bellows. In this way, Lethbridge could work for as long as six hours under water, although the rigid shape of the suit forced him to stay face-down all the time. The first of Lethbridge's many

dives was made in 1715, and by 1749 he reported having been down as far as 72 feet.

The prototype of the modern hard hat diving suit was invented in 1819 by Augustus Siebe and in 1837 developed into a full diving suit. This was a rubber watertight suit with a removable copper helmet. The helmet was fitted with intake and outlet valves and pressurized air was pumped into it from the deck of a ship. The Siebe diving suit enabled divers to go down to at least 300 feet. Using Siebe's device, the French zoologist Henri Milne-Edwards became the first scientist to explore the ocean floor in a diving suit. In 1844, he made a descent off the coast of Sicily and collected many specimens of marine life from the Mediterranean. Milne-Edwards repeated this exploit many times but, despite the growing availability of diving suits, few scientists followed his example.

In the mid-1800's, diving suits began to come into general use. But they still had one serious limitation. The diver continued to be dependent on a life line from the surface. In 1865, the Frenchmen Benoît Rouquayrol, a mining engineer, and Auguste Denayrouze, a naval officer, designed an apparatus to make the diver entirely self-sufficient. This was a metal canister, filled with compressed air, which the diver could carry on his back. The air was released by a regulator valve, while another valve removed the air breathed out by the diver.

Although a brilliant invention, the Rouquayrol-Denayrouze apparatus ran into a snag which marred its success. It was impossible at the time to make a canister which could withstand pressure at

Above: the diving helmet invented in 1819 by Augustus Siebe. It was supplied with pressurized air from the surface and formed the basis of the modern hard hat diving suit.
Below left: a boys' book illustration of the 1930's showing a helmeted diver recovering lost treasure.
Below right: a diver seeking pearls.

Above: a scene from the film based on Jules Verne's book *Twenty Thousand Leagues Under the Sea.* Verne modeled the breathing apparatus on a French invention of 1865. Below right: an illustration from the book, showing divers exploring the seabed.

great depths. Rouquayrol and Denayrouze were obliged to fall back on the use of a tube to pump air into the canister. It was to be another 78 years before divers could dispense entirely with their life lines.

But, if its original design was not wholly successful, the Rouquayrol and Denayrouze device became known all over the world through the writings of Jules Verne. The heroes of his book *Twenty Thousand Leagues Under the Sea* used this apparatus during their underwater excursions.

Jules Verne's book aroused a great deal of public interest toward the close of the 1800's. The adventures of divers and ways of living under the sea became a popular topic for writers, cartoonists, and inventors alike. At the same time, engineers and builders were starting to use diving suits and bells when constructing underwater foundations for bridges and harbors. Then, just as men began to feel more at home underwater, a series of seemingly inexplicable accidents again hindered the progress of diving.

Many of the underwater workers fell ill with a mysterious disease. They complained of severe pains in muscles and joints, of vomiting, fainting, and deafness. Some of them suffered from nervous disorders

Right: a cross-sectional diagram of David Bushnell's submarine *Turtle*. Its intrepid occupant was kept busy cranking two propellers as well as operating a rudder. Bushnell built the craft in 1776 so that it could be used to plant explosives on the hulls of enemy ships during the Revolutionary War in America. The powder charge was housed in a detachable section above the rudder. Below: Bushnell's submarine as it appeared from the outside.

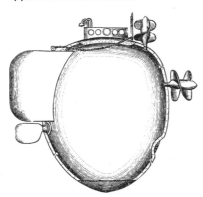

and paralysis. A number of men died suddenly. These men were victims of one of the greatest hazards of diving—the *bends*.

During the time a diver is under water, the nitrogen gas in the air that he breathes is continually being forced into his blood-stream and body tissues. The longer he stays down, and the deeper he goes, the more nitrogen enters his blood. As he returns to the surface, the reverse occurs. The nitrogen is returned to the lung surfaces and breathed out. But if the man ascends too rapidly, pressure drops quickly and the nitrogen trapped in his blood and tissues takes the form of bubbles. These bubbles are the cause of *decompression sickness*. If they block the blood flow to the brain or heart, they can cause permanent injury or death.

This phenomenon was first explained by French physiologist Paul Bert in 1870. Bert also explained that to avoid this decompression sickness a diver must rise to the surface very slowly so that the dissolved nitrogen comes out of his tissues gradually and bubbles do not form. Bert's work eventually led to the drawing up, in 1906, of a decompression table, setting out stages of ascent for divers. But decompression remained a major drawback by limiting the amount of time that divers can work under water and the depth to which they can go. For example, a dive to 600 feet, with 4 minutes spent at that depth, requires $11\frac{1}{2}$ hours of decompression.

This problem was very far from being solved in the early 1900's, when diving was called upon to assume a new and vital role—the rescue of submarine crews.

The development of the submarine had been going on side by side with that of the diving suit ever since the time of Leonardo da Vinci. Leonardo himself is said to have designed an underwater warship but

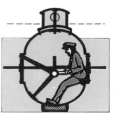

he kept his plans secret. "There is too much wickedness in the hearts of men," he wrote, "to justify my entrusting them with the secret of underwater navigation; they would not hesitate to use it to sow murder in the depths of the seas."

History was to prove him right. In 1776, David Bushnell built an underwater boat for use during the Revolutionary War in America. Bushnell named his submarine *Turtle* because of its shape. It was operated by a single man who hand-cranked two propellors—one to move the vessel forward or backward, the other to move it up or down—and controlled a rudder. Bushnell designed the submarine so that the operator could attach a powder charge to the hull of an enemy vessel. Using an explosive device, *Turtle* attacked a British

Above: diagrams showing two views of the Confederate submarine *Hunley*. It was built in 1863 and, like the *Turtle*, was hand-driven. A crew of eight operated a crankshaft connected directly to the propeller.

Below: an oil painting showing the *Hunley* during the Civil War. This submarine was the first to sink an enemy vessel, the Federal corvette *Housatonic*, on February 17, 1864, in the harbor at Charleston, S.C.

Above: the missile launching chambers on board the U.S. Navy nuclear submarine *George Washington.* Such submarines are important militarily but are of little use for exploration.

man-of-war in the harbor at New York, but was unable to sink it.

The design of the *Turtle* was improved on by American engineer Robert Fulton. In 1800, he built two submarines, the *Nautilus* and the *Mute.* Fulton believed that the submarine would help to end naval warfare and piracy, but neither of his vessels was ever used in war.

The first submarine to sink an enemy vessel was the Confederate *Hunley.* On February 17, 1864, during the Civil War, the *Hunley* sank the *Housatonic,* a Federal corvette that was blockading the harbor at Charleston, South Carolina.

The *Hunley,* like all the early submarines, was propelled by hand. Toward the end of the 1800's, inventors began to experiment with undersea vessels driven by compressed air, steam, and electricity. In 1898, John P. Holland, an Irish immigrant from New Jersey, built a submarine boat called the *Holland,* that used a gasoline engine on the surface and electricity underwater. The *Holland* was bought

by the U.S. Navy in 1900, becoming the Navy's first submarine.

Germany did not build a submarine until 1907, but it was her *Unterseeboten,* or U-boats, which demonstrated the power of submarines in naval warfare during World War I. The war speeded the development of improved submarines. It also revealed the urgency of discovering life-saving apparatus for their crews. Yet, as late as 1958, when America's nuclear-powered *Nautilus* had already made history by sailing under the North Pole, there were still no means of rescuing submarine crews from below 600 feet.

The submarine was designed primarily for military purposes, and as such contributed little to knowledge of the underwater world. The view from inside a submarine was limited and so were the depths to which it could descend. In order to reach and explore the greatest depths of the ocean, a new vehicle was needed. And it had to be a vehicle strong enough to withstand the tremendous pressures of the deep ocean.

Above: the world's first nuclear submarine *Nautilus.* This 3,000-ton ship made history in 1958, when it sailed under the ice at the North Pole charting a navigable route that could save 4,900 miles on an underwater voyage from Japan to Europe.

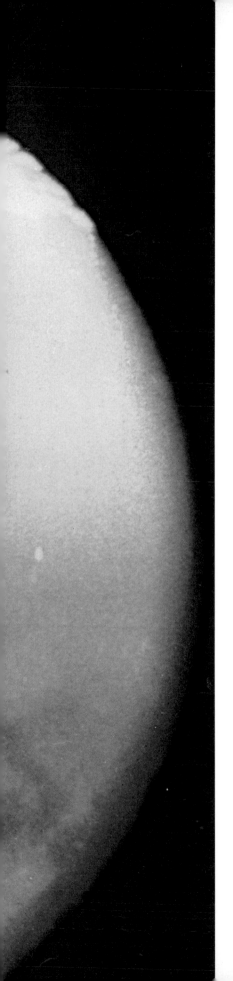

To the Bottom
5

Before 1930, no man had reached an undersea depth of more than about 600 feet. In that year, human beings were at last to penetrate the no man's land of the deep. Their record-breaking achievement was accomplished with the aid of an entirely new underwater vessel. This vessel was the brainchild of William Beebe, an American ornithologist turned marine zoologist.

Beebe had long been drawn to the oceans' depths in search of the living things that it hid. He had dived hundreds of times into the sea—always encased in a diving suit or diving helmet, which limited the depth to which he could go with safety. In the waters off Haiti, Beebe had slid down a rope to a bed of sand 63 feet below the surface. Peering through the glass of his copper helmet, he had walked across the sandy bottom until he had come to the edge of an undersea cliff. Beebe could see strange multicolored fish and other sea creatures darting through the waters below. In his book, *Half Mile Down,* Beebe writes, "As I peered down I realized I was looking toward a world of life almost as unknown as that of Mars or Venus ... a harvest [of life] which has served only to increase my desire actually to descend into this no man's zone."

Left: a modern bathysphere (depth sphere) being tested in shallow water. When tests are complete the bathysphere will be used to take men thousands of feet down into the ocean.

Right: the American naturalist and writer, William Beebe. With engineer Otis Barton and designer John Butler, Beebe set out to build a bathysphere to break the barrier of the deep. On June 6, 1930 Beebe and Barton descended in their steel sphere to a depth of 800 feet—200 feet deeper than man had ever gone before. It was the first of many successful dives.

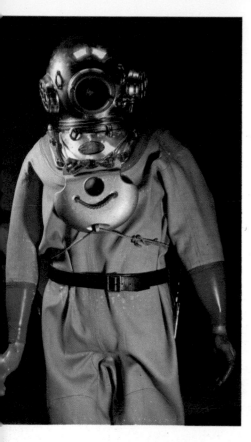

Above: a hard hat diving suit of the 1930's. William Beebe made many dives in a suit of this kind, but he was not content to explore only the shallow waters the suit was designed for. He wanted to go deeper.

Right: the bathysphere at the surface after a record-breaking dive of 2,510 feet, on August 11, 1934. Crew members of the stand-by barge *Ready* open the hatch to release William Beebe and Otis Barton from the sphere.

Below right: the bathysphere breaks surface as it is hauled from the sea.

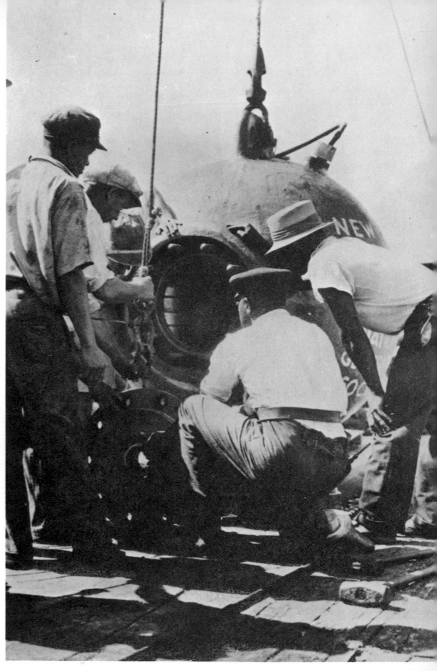

Beebe realized that he could not probe this zone in a diving suit. Eventually, water pressure would bar his way. There was only one way to break the pressure barrier—build a vessel strong enough to withstand hundreds of pounds of pressure per square inch and big enough to house a man or two in some comfort.

To accomplish this feat, Beebe turned to Otis Barton, an engineer who had already constructed a steel sphere for deep-sea exploration, and to John H. J. Butler, who had designed the vessel. Together, they reconstructed, modified, and outfitted the device. Beebe named it *bathysphere*—depth sphere.

The bathysphere was 4 feet 9 inches in diameter. Its steel walls were $1\frac{1}{4}$ inches thick and strong enough to withstand a pressure in excess of 1,500 pounds per square inch—equal to that at 3,400 feet below the sea's surface. It weighed 5,400 pounds. Unlike a military submarine, the sphere was equipped with portholes. These were made of three-inch-thick fused quartz, the strongest transparent material known to man and a substance that transmits all wavelengths of light. Each porthole was eight inches across. The bathysphere had a self-contained oxygen supply stored aboard in tanks. Moisture exhaled by the crew would be absorbed by trays of calcium chloride. Trays of soda lime, a mixture of calcium oxide and sodium hydroxide, would be used to remove from the atmosphere of the cabin the carbon dioxide exhaled by the men.

A communication hose would link the bathysphere to a mother ship at the surface. Cables in the hose would bring electricity to power instruments and a searchlight. And a telephone line would allow Beebe to give a "blow-by-blow" account of his plunge into the abyss. Finally, the bathysphere would be lowered on a steel cable, seven-eighths of an inch thick and 3,500 feet long. How deep would the bathysphere descend in its journeys into the sea?

The first answer to this question came on June 6, 1930. On that day, the bathysphere was transported into the warm waters off Bermuda on the deck of a huge barge, the *Ready*. The *Ready* itself was towed to sea by the ocean-going tug *Gladisfen*. During the previous few days, the bathysphere had made the same journey and had been lowered empty into the water a few times to test its seaworthiness. Now it was ready to carry human beings where none had gone before.

Shortly after noon, Beebe and Barton squirmed through the round 14-inch opening that served as a door for the bathysphere. On the outside, the deck crew maneuvered the 400-pound lid over the door opening. Huge bolts and nuts were joined to fasten the lid tightly. Beebe and Barton were sealed in their diving chamber.

The dangers for the two men were tremendous. Prisoners in their steel container, they could do nothing to counteract the jolting of the bathysphere. They were likely to be thrown about by sudden jerking of the cable as the mother ship pitched in the swell above them. At any moment, the strain on the cable could prove too great and the bathysphere plummet like a stone to the ocean bottom.

But danger was temporarily forgotten as Beebe looked out into the world beneath the waves. "At 50 feet," he noted, "I looked out at the brilliant bluish-green haze and could not realize that this was almost my limit in the diving helmet." At 100 feet, the light began to fade. As they sank farther, "motes of life" passed the portholes. Then, at 300 feet, an alarming trickle of water seeped from the door. "I wiped away the meandering stream," wrote Beebe, "and still it came. . . . I knew the door was solid enough . . . and I knew the inward pressure would increase with every foot of depth. So I gave the signal to descend quickly. . . ."

Two minutes later they passed 400 feet, then 500, and 600. The stream did not increase. At 700 feet, Beebe noted "only dead men

Above: William Beebe clambers out of the bathysphere after descending to 2,510 feet in the sea off Bermuda. This dive was one of a series that he and Otis Barton made during 1934.

Above: a deep-sea red prawn (*Systellaspis sp.*) holding a three-inch-long minnow of the deep *(Cyclothone sp.)*. Both these creatures have light-producing organs and were seen frequently by William Beebe as he descended to depths below the level of 1,000 feet.

have sunk below this." Halting in the journey downward, he described the water outside as "an indefinable translucent blue quite unlike anything I have ever seen in the upper world." As he switched on the searchlight he saw strange fishes darting in and out of the beam. When the light was turned off, luminescent creatures seemed to fill the water with eerie specks of light.

The bathysphere sank deeper into the sea and "the twilight deepened . . . from dark blue to blacker blue." "800 feet," came the call from the surface crew. "Stop!" replied Beebe. The bathysphere had reached "bottom" for this maiden voyage.

During the next 4 years, Beebe and Barton made more than 30 dives in the bathysphere. But none was more dramatic than dive number 32 near Bermuda on August 15, 1934. During the early series of dives in 1930, the two undersea explorers had plumbed the sea to a depth of 1,426 feet. Then the bathysphere had been put on display at the Century of Progress Exposition at Chicago in 1933. Following the exposition, the National Geographic Society offered to sponsor a new series of deep dives. Number 32 was the highlight of Beebe and Barton's final series of probes. It broke all their previous records both for the depth reached and for the number of living creatures observed.

In his account of the dive, Beebe wrote, "Surprises came at every

few feet . . . the mass of life was totally unexpected, the sum total of creatures seen unbelievable. At 1,000 feet I distinctly saw a shrimp outlined and distinguished several of its pale greenish lights. . . . Large Melanostomiatid dragon-fish with their glowing porthole lights showed themselves now and then [and] we had frequent glimpses of small opalescent copepods [small crustaceans], appropriately called *Sapphirina,* which renewed for us all the spectrum of the sunlight.

"At 1,680 feet. . . . I saw some creature, several inches long, dart toward the window, turn sideways and—explode. . . . At the flash, which was so strong that it illumined my face and the inner sill of the window, I saw the great red shrimp and the outpouring fluid of flame." (This is a defense process of certain shrimps, which confuses predators in a similar way to the ink clouds produced by a threatened octopus.)

"At 1,800 feet I saw a small fish with illumined teeth . . . and ten feet below this my favorite sea-dragons, *Lamprotoxus,* appeared, they of the shining green bow. Only sixteen of these fish have ever been taken. . . . The record size is about eight inches, while here before me were four individuals all more than twice that length, and very probably representing a new species. . . . At 2,450 a very large, dim, but not indistinct outline came into view for a fraction of a

Above: a saber-toothed viperfish (*Chauliodus* sp.—up to 10 inches long). Beebe reported seeing 7 of them at 1,700 feet. This fish has light-producing organs on its body and inside its mouth. As it opens its mouth the lit interior may act as a lure to crustaceans and small fishes.

Above: Beebe and Barton after their spectacular dive in the bathysphere on August 15, 1934. On that day they dived to a depth of 3,028 feet.

Below: a deep-sea angler *(Onsirodes carlsbergi).* At a depth of 1,900 feet William Beebe reported seeing "one of the true giant female anglerfish . . . a full two feet in length. . . ."

second, and at 2,500 a delicately illumined ctenophore jelly throbbed past. Without warning, the large fish returned and this time I saw its complete, shadow-like contour. . . . Twenty feet is the least possible estimate I can give to its full length. . . . For the majority of the 'size-conscious' human race this MARINE MONSTER would, I suppose, be the supreme sight of the expedition. . . . What this creature was I cannot say. A first, and more reasonable guess would be a small whale or blackfish. We know that whales have a special chemical adjustment of the blood which makes it possible for them to dive a mile or more, and come up without getting the 'bends.' So this paltry depth of 2,450 feet would be nothing for any similarly equipped cetacean.

"Soon after [Barton] saw the first living *Stylophthalmus* ever seen by man. . . . This is one of the most remarkable of deep-sea fish, with eyes on the ends of long, periscope stalks, almost one-third as long as the entire body. . . .

"At 11.12 A.M., we came to rest gently at 3,000 feet, and I knew that this was my ultimate floor; the cable on the winch was very near its end. . . . The water . . . seemed to show as blacker than black. It seemed as if all future nights in the upper world must be considered only relative degrees of twilight. I could never again use the word BLACK with any conviction.

"Now and then I felt a slight vibration and an apparent slacking off of the cable. Word came that a cross swell had arisen, and when the full weight of bathysphere and cable came upon the winch, Captain Sylvester let out a few inches to ease the strain. There were only about a dozen turns of cable left upon the reel. . . . We were swinging at 3,028 feet, and, Would we come up?"

They did come up—from the greatest depth reached by a living

human being. But the record was not to go long unchallenged.

While Beebe and Barton were making their historic journeys downward, two other pioneers were traveling upward into the unexplored heights of the stratosphere (the second layer of the atmosphere). In 1932, the Swiss physicist Auguste Piccard and his assistant Max Cosyns reached a record height of 53,139 feet in a balloon and gondola of Piccard's invention. (A gondola is an airtight ball in which balloonists travel instead of a basket when exploring the upper atmosphere.)

To Piccard, who had been intrigued by the undersea depths since boyhood, there were clear analogies between the bathysphere and the balloon "In both cases," he wrote, "there is a danger of the cable breaking, with the difference, however, that the aeronaut . . . cannot help wishing: 'If only this rope would break, what a fine trip in a free balloon we should have.' Very much to the contrary, the oceanographer, shut up in his tight cabin, is haunted by the terrifying idea that the cable may break. But can we do without the cable?"

Piccard had long dreamed of an underwater vessel that would need no lifeline to descend to the greatest depths of the ocean. Not long after his balloon ascent, he set about designing just such a device—a bathyscaph, or deep-sea ship. Based on the principle of the balloon, the bathyscaph was intended to float free, up or down. It consisted of two major parts, a cabin and a float. The cabin was very much like a gondola and the float was like a balloon.

The cabin, built to withstand the crushing pressures of the deep ocean, was a steel sphere weighing 10 tons. Its internal diameter was $6\frac{3}{8}$ feet and its wall was 3.54 inches thick (5.91 inches thick around the 2 cone-shaped portholes). The portholes were made of a newly developed, shatterproof plastic called Plexiglass, which will

Above: the Swiss physicist Auguste Piccard (right) at a balloon meeting with King Leopold of Belgium. Piccard later used the principle of the balloon in the design of his first bathyscaph or deep-sea submersible. Below: Piccard climbing into the steel gondola before a balloon ascent to over 51,000 feet, in 1931.

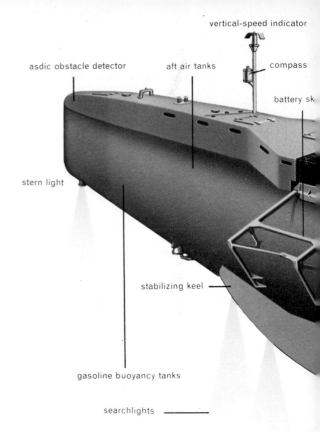

asdic obstacle detector   aft air tanks   vertical-speed indicator

compass

battery sk

stern light

stabilizing keel

gasoline buoyancy tanks

searchlights

Above: a test forging of the *Trieste* gondola
(the sphere in which the explorers traveled)
at the Deutsches Museum, Munich.

not crack under pressure. Each porthole was 5.91 inches thick, with
a 3.94-inch inside diameter, and a 15.75-inch outside diameter.

But how could the descent and ascent of such a sphere be accomplished and, more important, delicately controlled? If the sphere hit
bottom too hard, it would be destroyed and its daring passengers
killed. To solve this vital problem, Piccard turned to his experience
in the air. In simple terms, a balloon rises because it is lighter than
the air it displaces. Piccard's balloon had been filled with 100,000
cubic feet of lighter-than-air hydrogen gas. This was enough
hydrogen to make the total volume of the balloon and gondola
lighter in weight than the same volume of air surrounding it. The
balloon would rise until the weights of these two volumes became
equal. Because the atmosphere becomes thinner—less dense—with
altitude, there is a limit to the height a balloon can soar. To some
extent, that limit depends on the volume of hydrogen in the balloon.
So there is only one way to bring such a balloon back down—vent
hydrogen from it.

Piccard could not fill the float of the bathyscaph with hydrogen
or any other gas. Gases are too compressible. As the bathyscaph sank
into the ocean depths, increasing pressure would cave in the walls
of the float. (The float had to be lighter than water and its walls,
therefore, thin-skinned.) After a long search, and a number of
experiments, Piccard decided to use common gasoline as the
buoyant material in the float. Gasoline had a number of advantages
that suited Piccard's needs. It was less dense than water, it was

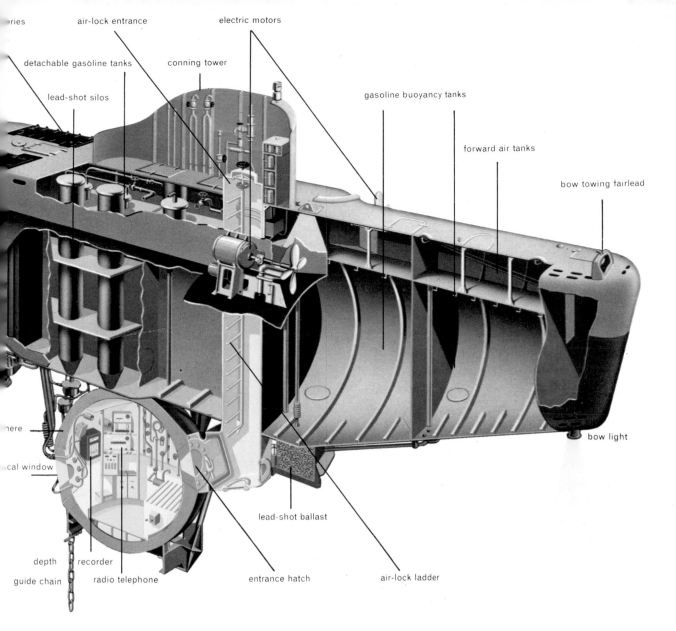

ries

air-lock entrance

electric motors

detachable gasoline tanks

conning tower

lead-shot silos

gasoline buoyancy tanks

forward air tanks

bow towing fairlead

here

cal window

bow light

lead-shot ballast

depth   recorder

guide chain   radio telephone

entrance hatch

air-lock ladder

Above: A cutaway diagram of the bathyscaph *FNRS 3*. Based on Piccard's designs, it was built by the French Navy at Toulon. Early in 1954, Georges Houot and Pierre Willm took the *FNRS 3* to the record depth of 13,287 feet below the surface.

Left: Professor Auguste Piccard and his son, Jacques. In 1951, while the *FNRS 3* was still under construction, Piccard received the money to build a new bathyscaph. It was to be named the *Trieste*.

relatively non-compressible, and it was insoluble in water.

In addition to the gasoline tanks, the float was equipped with two air tanks—one at each end of the float—and two ballast hoppers filled with tons of iron shot. When each tank was filled with its appropriate material, the bathyscaph would float on the surface of the sea. However, if water were allowed to enter the air tanks, the bathyscaph would begin a gentle plunge downward. If Piccard wanted to hover, he would drop a small amount of shot. If he wanted to return to the surface, he would drop more shot. The shot was held in the hoppers by the pull of electromagnets situated on top of the float. When the electric current in the electromagnets was cut off from inside the cabin by the pilot—or by some accident—shot would fall out of the hoppers and the craft would rise. To increase the rate of descent, gasoline could be jettisoned from one of the six gasoline tanks.

Left: The Piccards' bathyscaph *Trieste* suspended above the sea. The observation gondola, in which the two men sat, can clearly be seen below the gasoline-filled float.

Pressure was no threat to the float's thin walls because the float was open to the sea. Water could flow freely into the bottom of it. The pressure on the inside of the float always equaled that on the outside. Moreover, because gasoline is less dense than water, the gasoline sat on top of the entering water and could not leak from the openings through which the water entered.

This, essentially, was the ingenious device that was hoisted by winch from the hold of the Belgian cargo ship, *Scaldis,* shortly before

Below: the bathyscaph *Trieste* at the start of a dive. The *Trieste* underwent her first sea trials at Castellammare di Stabia, near Naples in Italy, early in August, 1953.

1 P.M., on November 3, 1948. The *Scaldis,* accompanied by the French oceanographic vessel *Élie Monnier,* had sought out a spot over 4,600 feet deep in the Atlantic Ocean off Dakar (now the capital of Senegal), near the Cape Verde Islands. Here, Auguste Piccard and his 26-year-old son Jacques would put the bathyscaph through its first test dive into the deep ocean. For this unmanned dive, a timing device aboard the bathyscaph would automatically release the iron shot. Piccard had set the depth gauge for 770 fathoms (4,620 feet) and the bathyscaph had 40 minutes to reach it.

It was Jacques Piccard who opened the air tanks of the float while the bathyscaph bobbed on the choppy Atlantic waters. As the seawater rushed in and the bathyscaph began to sink, he jumped clear and joined his father on board the *Scaldis.* The minutes dragged as a tense band of scientist-oceanographers waited anxiously for the moment the bathyscaph would surface again. Clinging to masts and funnels, sailors of the *Scaldis* and *Élie Monnier* joined the watch. At last, a cry came from the *Scaldis. "Le voila!* There it is!"—and the bathyscaph shot through the surface.

Later, readings of the depth gauge in the cabin revealed that the bathyscaph had plumbed the sea to a depth of 4,554 feet. With the exception of a few drops of water in the cabin and a lost radar antenna it had returned in good shape. If it had carried men in its cabin, they would have glimpsed sights never before seen, and returned to tell about them. But the first manned descent was not to be made that day nor for many days. The drops of water, the lost antenna, and an inability to pump off the bathyscaph's gasoline ended the Dakar dives. The first two accidents had to be checked out before men could be allowed to dive in the vehicle. And the gasoline had to be piped into the hold of the *Scaldis* before the bathyscaph could be hoisted aboard for inspection. Unfortunately, through damage or some other inexplicable reason, the hose used to siphon gasoline from the bathyscaph to the *Scaldis* could not be connected properly to the mother ship. Auguste Piccard had no choice but to order the 6,600 gallons of gasoline to be dumped into the sea.

Now came months of frustration. Money was needed for a new supply of gasoline and, more important, for improvements in the design of the bathyscaph. Jacques Piccard scoured Europe for funds. Years passed, but slowly money began to trickle in. Technical problems were eventually solved and a new bathyscaph was constructed. Piccard named it the *Trieste* for the Adriatic seaport where it was built.

The *Trieste* was launched on August 1, 1953, at Castellammare di Stabia, near Naples, in southern Italy. During the next few days preparations were made for shallow test dives in the harbor of Castellammare. Then, on August 11, 1953, Auguste and Jacques Piccard crawled into the cabin of the *Trieste,* sealed themselves in, flooded the float's air tanks, and slowly settled to the bottom of the harbor 26 feet below the surface. After jettisoning iron shot, the *Trieste* gently rose to the surface. The first dive had been successful.

Below: U.S. Navy scientist A. B. Rechnitzer (left) and Jacques Piccard on board *Trieste.* On November 15, 1959 Piccard piloted *Trieste* to a depth of 18,150 feet south of the island of Guam in the Pacific Ocean. It was one of the dives made after *Trieste* was bought by the United States Office of Naval Research.

During the years that followed, 64 more dives were logged by the *Trieste*. The Piccards, father and son, rode down 3,540 feet near the island of Capri in the Bay of Naples, and 10,300 feet near the island of Ponza in the Tyrrhenian Sea, west of Naples. In 1958, the *Trieste* was purchased by the United States Office of Naval Research. From then on, Jacques Piccard teamed up with U.S. scientists, engineers, and naval officers to dive to 12,110 feet south of Ponza, to 18,150 feet south of the island of Guam in the Pacific Ocean, and to 23,000 feet, also off Guam. But each of these record-breaking dives was really a dress rehearsal for the most daring dive of all—the attempt to touch down on the bottom of the deepest spot on earth, the Challenger Deep.

The bottom of the Challenger Deep lies 36,198 feet down in the Mariana Trench, southwest of the island of Guam in the Pacific. It was named for H.M.S. *Challenger II,* the British oceanographic research vessel which discovered it in 1951.

Nine years later, on January 23, 1960, the bathyscaph *Trieste* rode the Pacific waters directly above "the basement of the world." It had been towed there by the U.S. naval ship *Wandank.* Aboard the *Wandank* and its destroyer escort, U.S.S. *Lewis,* were Jacques Piccard, his American colleague Donald Walsh of the U.S. Navy, and a team of scientific and naval experts. Walsh had joined Piccard in five previous dives, including number 64 (to 23,000 feet). On this day, he and Piccard would try to reach bottom at the greatest depth ever recorded.

The sea was dangerously rough. Twenty-five-foot waves smashed against the *Wandank* as Piccard jumped aboard a rubber raft and made for the *Trieste.* Once he had reached the bathyscaph, he found to his dismay that the surface telephone, used to communicate from the cabin to the surface just before diving, had been washed overboard. The tachometer, which measures rate of descent, had also been put out of commission by the battering waves, and the dive was in fact made without it.

Walsh joined Piccard aboard the *Trieste* and the two men debated whether or not to risk the dive. Piccard proposed that if all were well in the cabin, they should go ahead. He climbed down to the cabin from the conning tower atop the float and switched on the electromagnets. They worked—and so did the other instruments. The dive was on!

At 8:23 A.M. the *Trieste* began to drift slowly downward. Then at 340 feet, it seemed to go into reverse, bouncing up a few yards. It had struck a common undersea obstacle—a *thermocline,* a body of water separating the warm surface water from the colder depths. The thermocline was colder and therefore denser than the water through which the *Trieste* had just descended—sufficiently dense to buoy the *Trieste* upward. Piccard opened the valve of one of the gasoline tanks. The bathyscaph, heavier now, broke through the thermocline. Down it went at an average speed of four inches per second. By 9 A.M., the *Trieste* had only plunged 800 feet into the sea. At this

Left: the bathyscaph *Trieste* in harbor. After completing many dives— and setting many new depth records— *Trieste* was made ready for the deepest dive of all—into the darkness over 35,000 feet down in the Challenger Deep, southwest of the island of Guam in the Pacific Ocean. Below: life on the soft mud of the deep ocean floor, photographed with a deep-sea camera. Two brittle stars (a type of echinoderm) and a deep-sea fish can clearly be seen.

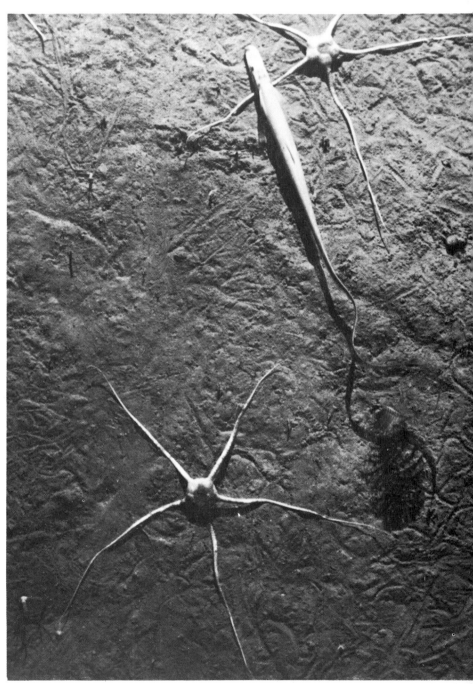

rate, it would not have time to reach bottom and resurface before night, and Piccard would not chance a night-surfacing in calm seas, let alone turbulent ones. More gasoline was vented and the *Trieste* accelerated to three feet per second.

At 1,000 feet, Piccard turned on the outside floodlight. Plankton seemed to swarm upward like a blizzard defying gravity. With the floodlight and cabin lights off, the sea appeared a dark gray-black. At 2,400 feet, total blackness engulfed the bathyscaph. The cold water outside sent a chill into the cabin. Walsh and Piccard changed their wet clothing which had been soaked with seawater when they had boarded the *Trieste,* and the two explorers munched chocolate bars for "lunch."

At 4,200 feet the sphere sprang a leak but sealed itself shortly afterward. Another leak at 18,000 feet abruptly self-sealed. Any slight crack in the sphere was quickly closed by the enormous pressure building up on its surface.

The *Trieste* soon dipped below 23,000 feet, topping record dive number 64. No men had been deeper. At 11:30 A.M., the depth gauge read 27,000 feet. How far to the bottom? Piccard wondered. He dropped some iron shot from the float's hoppers and the *Trieste* sank more slowly. Soon after passing the 29,000-foot level, it was "as deep under the sea as Mount Everest is high above it." The floodlights probed crystal-clear water with no sign of life, large or small. Piccard dumped more iron ballast. The bathyscaph slowed to a descent of one foot per second.

At noon, the *Trieste* reached 31,000 feet. It was rapidly nearing a depth whose pressure had never been tested on its sphere or float. Piccard switched on the echo sounder, which was designed to detect the bottom if it were 600 feet or nearer to the descending bathyscaph. No echo returned. The *Trieste* continued its journey downward— soon it passed 35,000 feet. The men anxiously watched the echo sounder. At last, they received an echo from the bottom—only 252 feet below.

Minutes later, Walsh had counted down the last few fathoms and the *Trieste* touched the ocean floor. Its official depth, given after slight correction of the depth gauges, was 35,800 feet, deeper than any man had yet penetrated and only 398 feet short of the deepest spot in the whole ocean bed.

Two hundred thousand tons of pressure gripped the *Trieste* as the two men scanned the ocean floor. Piccard spotted a flatfish lying on the tan-colored ooze. It was about one foot long and six inches across. Two round eyes protruded from the top of its head. In the enormous pressure of the abyss, there was life.

The two explorers stayed for 20 minutes on the seabed. At 4:56 P.M. they returned to the surface in perfect health. Their historic dive had broken every undersea record. They had reached one of the deepest spots in the earth's oceans and in doing so had proved that man could explore the bottom of the sea and return safely to the surface.

Above: the crew sitting on *Trieste* at sea. On January 23, 1960, rough conditions in the Pacific Ocean threatened to stop Jacques Piccard and Donald Walsh from starting their descent into Challenger Deep. However, all was well in the cabin of the *Trieste* and the dive did take place as originally planned.

Right: a diagrammatic view of the Mariana Trench, showing *Trieste*'s dive into Challenger Deep (H). The diagram also shows: (A) the thermocline; (B) the point where light fades out; (C) the deepest point to which a whale can dive; (D) Beebe's dive in the bathysphere; (E) Barton's benthoscope; (F) the *FNRS 3* bathyscaph, and (G) a deep-sea camera. Pressure increases with depth, and the *Trieste,* resting on the bottom 35,800 feet below the surface, is gripped by 200,000 tons of pressure, or 8 tons per square inch.

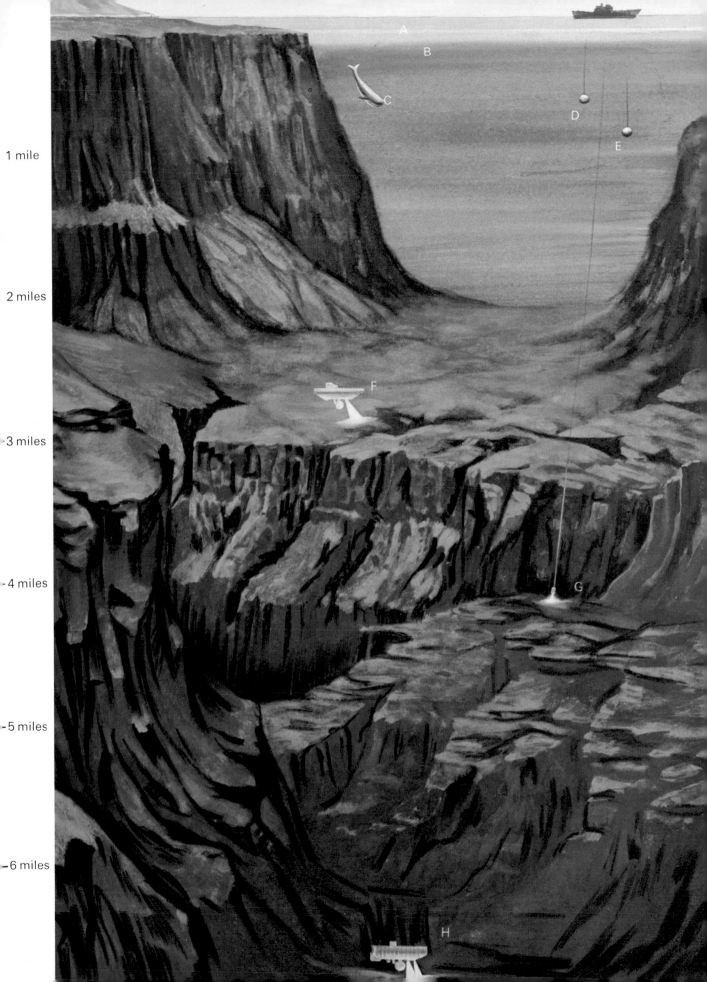

1 mile

2 miles

3 miles

4 miles

5 miles

6 miles

Above: Benjamin Franklin. In 1770, when Franklin was deputy postmaster general, mail ships from Britain were held up by the Gulf Stream on their way to Massachusetts. Franklin ordered the preparation of a chart showing the current's course, but it was some time before the British captains took advantage of his advice.

Right: the *Ben Franklin,* a research submersible designed by Jacques Piccard and built and operated by the Grumman Aerospace Corporation. On July 14, 1969, the *Ben Franklin* set out on a journey of undersea discovery in the Gulf Stream.

A River in the Sea

6

On July 14, 1969, two days before the liftoff of Apollo 11, crowds of people began gathering near Cape Kennedy to watch the final preparations for man's first landing on the moon. Probably only a handful of these excited onlookers were aware that, a mere 130 miles away, at West Palm Beach, Florida, another historic launching was about to take place. This was the launching of an experimental craft designed by Jacques Piccard and called the *Ben Franklin*. Like Apollo 11, this vehicle was crammed with scientific instruments for a voyage into a strange and hostile environment. But, unlike Apollo 11, the *Ben Franklin*'s voyage was to begin, rather than end, with a

splashdown. This craft, a brand new type of underwater research vessel, was about to undertake the first undersea exploration of the mighty Gulf Stream.

The Gulf Stream is the second largest ocean current in the world. (Only the Antarctic Circumpolar Current is greater.) About 50 miles wide and 3,000 feet deep, it flows from the Gulf of Mexico north-eastward through the Atlantic toward the shores of Europe. For centuries, the warm Gulf Stream has intrigued and puzzled both sailors and scientists. But only during the past 200 years has any real progress been made toward understanding the origin of this river

Above: a painting showing the danger and skill involved in whaling in small boats in the mid-1800's. It was during their whaling expeditions that American sea captains became familiar with the Gulf Stream.

in the sea, and learning more about its movement and behavior.

Man's first recorded encounter with the Gulf Stream took place in 1513. In that year, Juan Ponce de León, the Spanish conqueror who explored Florida, was sailing southward down the peninsula's east coast. Suddenly, his three ships, which had been making steady progress before a fresh wind, were brought to a complete halt by a mysterious northward-pushing current. Describing this strange event in his log, Ponce de León wrote, "We held to the south [but] could make no headway. . . . Eventually we had to recognize that despite inflated sails we were being driven backward and that the current was stronger than the wind. Two of the ships, which were somewhat nearer the coast, were able to cast anchor, but the current was so strong that it broke the ropes!"

In the two centuries that followed Ponce de León's voyage, many sailors and explorers crossed the Gulf Stream, bucked its currents or rode with them. These encounters with the river in the sea produced numerous conjectures, theories, and tall tales about its existence and curious characteristics. But it was not until 1770 that the first chart of the great current's path through the Atlantic was made. In that year a complaint was lodged by the customs officials in Boston concerning slow delivery of mail from England to the Massachusetts colony. It seemed to the Bostonians that the mail packets were dawdling on their way across the Atlantic from the Old World to the

Below: Benjamin Franklin's chart of the Gulf Stream. As deputy postmaster general of the colonies, Franklin advised captains of British mail packets sailing west, "Don't fight the Gulf Stream."

New. Why, they wanted to know, should it take the English mail ships two weeks longer to cross the ocean than it did the American vessels going the other way? Benjamin Franklin, who was then deputy postmaster general, put the problem before Nantucket whaling captain Timothy Folger.

Folger's answer was simple—the English captains were not sufficiently familiar with the Gulf Stream. "We are well acquainted with the stream," Folger explained, "because in our pursuit of whales, which keep to the sides of it but are not met within it, we run along the side and frequently cross it to change our side. . . . In crossing it [we] have sometimes met and spoken with those packets who were in the middle of it and stemming it. We have informed them that they were stemming a current that was against them to the value of three miles an hour and advised them to cross it, but they were too wise to be counseled by simple American fishermen."

At Franklin's request, Folger marked out the course of the Gulf Stream on an Atlantic chart. Franklin sent the chart to England, but the English captains are reported to have "slighted it" for some time before realizing its value and following Folger's advice.

On and off during the next 160 years oceanographers ventured into the warm waters of the Gulf Stream in search of its secrets. Using submersible thermometers and sampling bottles, they gathered information about its temperature and density at various depths. *Drift bottles* were labeled and tossed overboard to be recovered later in unexpected places, revealing new eddies and arms of the stream. From this and other information, scientists determined that the stream is created by a combination of winds and currents. Trade winds cause a westerly flow of water where the Atlantic Ocean crosses the equator. This flow passes through the narrow Yucatán Channel into the Gulf of Mexico, where it gathers momentum and flows out through the Straits of Florida.

In 1930, the study of the Gulf Stream took an important step forward with the founding of the Woods Hole Oceanographic Institution at Cape Cod, Massachusetts. One of the primary objectives of the scientists at Woods Hole was to probe the waters of the Gulf Stream. To carry out this mission, they purchased the ketch *Atlantis*. A ship of 460 tons, the *Atlantis* was outfitted with the most modern scientific equipment. and could carry a crew of 25 sailors and oceanographers. Over the next few years, it logged more than half a

Above: the 460-ton ketch *Atlantis*. Launched in 1931, it was outfitted with the latest scientific equipment and used by the newly-formed Woods Hole Oceanographic Institution for research on the Gulf Stream.

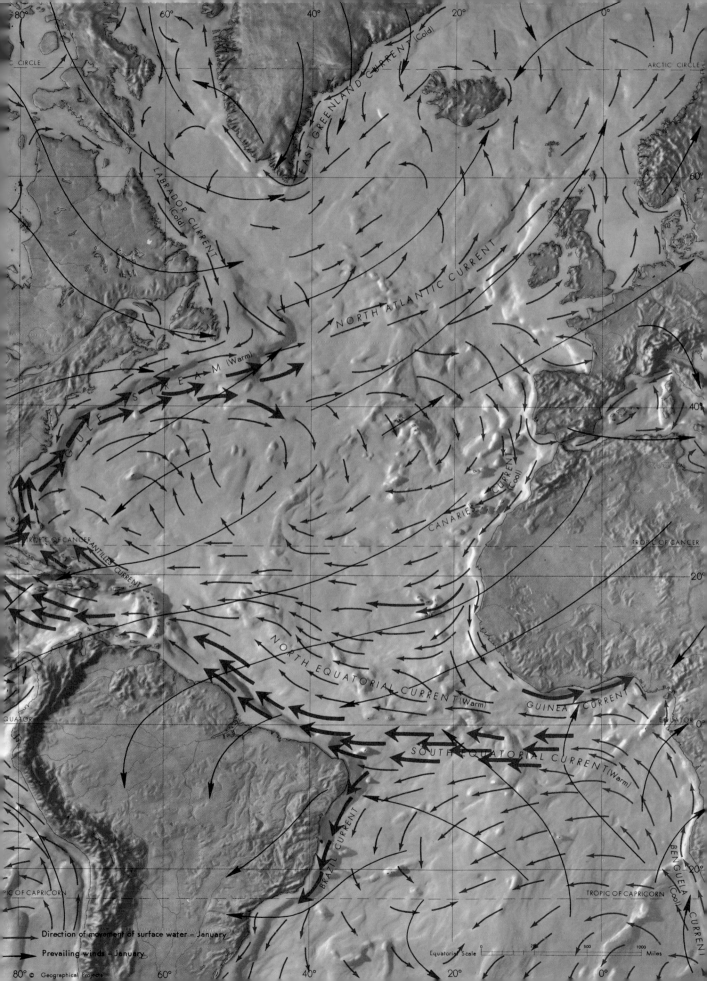

ARCTIC CIRCLE

ARCTIC CIRCLE

EAST GREENLAND CURRENT (Cold)

LABRADOR CURRENT (Cold)

NORTH ATLANTIC CURRENT

GULF S T R E A M (Warm)

CANARIES CURRENT (Cool)

TROPIC OF CANCER

TROPIC OF CANCER

ANTILLES CURRENT

NORTH EQUATORIAL CURRENT (Warm)

GUINEA CURRENT

EQUATOR

EQUATOR

SOUTH EQUATORIAL CURRENT (Warm)

BRAZIL CURRENT

BENGUELA CURRENT (Cool)

TROPIC OF CAPRICORN

TROPIC OF CAPRICORN

Direction of movement of surface water – January

Prevailing winds – January

Equatorial Scale   500   1000   Miles

80° © Geographical Projects

million miles as it sailed to and fro along the stream's pathways.

In the course of their many exploratory missions, the Woods Hole scientists were able to amass an extraordinary amount of information about the Gulf Stream, particularly in regard to the volume of water it carries. Using a variety of sophisticated submersible equipment, they were able to determine that, as the stream heads out into the Atlantic south of New England, it is transporting something like 150-million cubic meters of water a second. (The mightiest river in the United States, the Mississippi, disgorges only about a thousandth

Left: the crew of the mesoscaph *Ben Franklin* on deck on July 14, 1969, before the start of their journey in the Gulf Stream. Jacques Piccard is at the right of the group.

Right: a cutaway diagram of the *Ben Franklin*. The 49-foot-long mesoscaph was packed with scientific instruments, including cameras to record the crew's reaction to their 30 days of confinement. During the journey the crew made observations of current velocity, temperature, salinity, animal life, and other characteristics of the Gulf Stream.

1. Ballast tank
2. Scientific control center
3. View port
4. CO_2 filter
5. Hydrophones
6. Surface radio antenna
7. T.V. camera
8. Surface lookout
9. Conning tower
10. Control console
11. Hatch
12. View ports
13. Instrument package, cameras, etc.
14. Propulsion motor
15. Forward oxygen tank
16. Side-looking sonar
17. Batteries
18. Ballast tank
19. Biological sampler
20. Stern oxygen tank

of this amount of water at its mouth.) This measurement of the Gulf Stream's tremendous volume becomes even more amazing when we realize that at its source it is carrying only about 30 million cubic meters of water per second. Where the extra water comes from remains a mystery.

This was in part the state of knowledge about the Gulf Stream when the *Ben Franklin* glided out of West Palm Beach on July 14, 1969. A *mesoscaph*—middle boat—it was designed to explore the ocean at depths up to 2,000 feet. Its 49-foot-long hull was filled with scientific and navigational instruments, as well as all the necessary living facilities for its six-man crew.

The *Ben Franklin*'s mission, known as the Gulf Stream Drift Mission, was conceived by Jacques Piccard in 1965. Unlike the bathyscaph project, this idea found almost instant support. The Grumman Aerospace Corporation financed and drew up final plans for the construction of the *Ben Franklin*. The Oceanographic Office of the Department of the Navy (NAVOCEANO) provided about 2,300 pounds of scientific instrumentation for the vessel, and two oceanographers as crew members, as well as a surface support vessel. The National Aeronautics and Space Administration (NASA) supplied an observer to go along on the drift to evaluate crew

Right: part of the control room on board *Ben Franklin*. The elaborate control systems were vital to the mission. During the journey the submersible was put through many delicate maneuvers, including several trips to investigate the sea bottom at depths of 1,200 to 2,000 feet.

reactions during the month-long voyage. In return, Piccard and his crew were to perform three groups of experiments.

For Grumman, the men of the mesoscaph would search for plankton and minerals in the Gulf Stream waters. NAVOCEANO experiments would cover a wide variety of subjects including periodic measurements of current speed, available light, gravity, magnetic field, natural ocean noises, bottom structure, and animal life. NASA was primarily interested in how men would react to long confinement and isolation—conditions that would exist in space stations. For NASA's experiments, tape recordings were made of crew conversations during meals. Three fixed cameras inside the vessel snapped photographs every two minutes to record the behavior of the submariners, and each member of the crew was required to keep a diary.

In the course of the 30-day 11-hour voyage, the crew of the *Ben Franklin* experienced periods of extreme tension, and occasionally found their enforced captivity and cramped quarters almost unendurable. For the most part, however, their morale remained high and, at the conclusion of their mission, they could point with pride to a record number of achievements.

The Gulf Stream had carried the *Ben Franklin* 1,444 nautical miles to a spot in the North Atlantic 360 miles south of Nova Scotia. Though it had drifted at an average depth of 650 feet, it had made 10 trips to depths between 1,200 and 1,800 feet—5 of which were ocean bottom surveys. The crew had made more than 900,000 temperature, salinity, and sound velocity (speed) measurements—the latter to determine water density. Sixty thousand photographs of the crew were snapped as the men went about their routine duties. NASA psychologists spent many hours poring over these.

Surprisingly, crew members peering out of the *Ben Franklin*'s

Below: the eastern coastline of the United States, showing the route of the mesoscaph *Ben Franklin* in 1969. The *Ben Franklin* explored the Gulf Stream at depths up to 2,000 feet from West Palm Beach, Florida, to a point 360 miles south of Nova Scotia. Its expedition was known as the Gulf Stream Drift Mission.

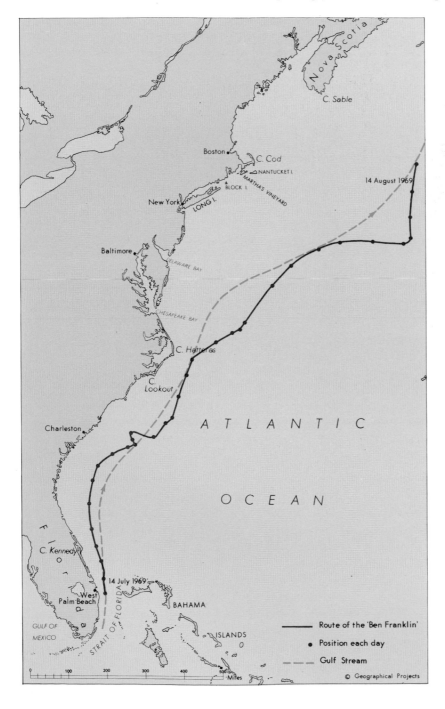

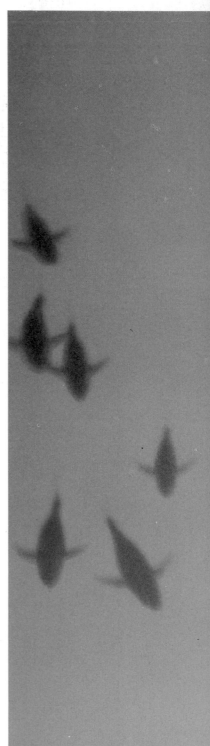

observation ports had spotted relatively few fish and scarcely any plankton in the Gulf Stream. Probes of the stream with pulses of sound had failed to detect a *deep scattering layer* common to many parts of the seas. (The deep scattering layer refers to the dense blanket of marine organisms which reflect back sounds as scattered markings on an echo sounder. Such layers of living things are generally found in depths between 600 and 2,400 feet. And they are usually between 150 and 600 feet thick.) As far as life was concerned, the Gulf Stream was a watery waste.

Below: a photograph taken from the *Ben Franklin* shows a school of tuna swimming overhead. This was an unusual sight for the crew, who saw few signs of life in the Gulf Stream.

Left: Jacques-Yves Cousteau. In 1943 Cousteau tested the first Aqua-Lung, a revolutionary underwater breathing device that enabled divers to explore beneath the sea unencumbered by air-lines linking them to the surface.

Below: a diver wearing a modern Aqua-Lung is able to match the agility of a fish in its natural environment.

Man Lives in the Sea
7

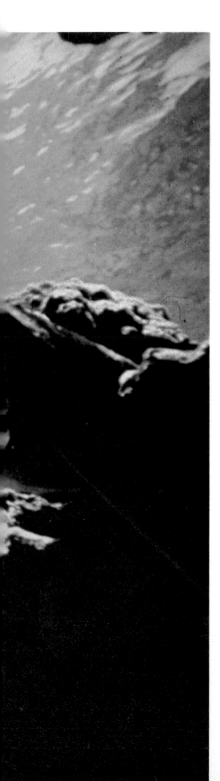

In the summer of 1943, at the height of World War II, a man waded into the Mediterranean off the coast of southern France. The cove in which he waded was sheltered from the eyes of the occupying Axis troops and he had chosen it for that reason. Strapped to his back was a strange apparatus that the enemy armies would have considered priceless.

The man's name was Jacques-Yves Cousteau. The equipment on his back was the first Aqua-Lung—a device that was to revolutionize the exploration of the oceans and open their depths to the individual diver. The Aqua-Lung transformed the old-fashioned diver, with his cumbersome copper helmet and weighted boots, into a *man-fish* who could glide at will beneath the water.

Cousteau and several others had begun skin diving for sport in the Mediterranean in the years before World War II. The small band of underwater pioneers had developed fins, goggles, and fishing spears, and had experimented with several forms of underwater "lungs." But the primitive lungs used pure oxygen and were dangerous at high pressures. On two occasions Cousteau almost drowned when using one of these early devices. What was needed, Cousteau realized, was some kind of compressed-air lung that would automatically release air at the proper pressure on demand from the diver.

The war scattered Cousteau's group of divers. After the signing of the Franco-German armistice in 1940, they came together again. But now their underwater projects had to be carried out in the strictest secrecy. France was occupied by Axis troops, and the French government cooperated to a great extent with the occupying forces, although many Frenchmen, both inside and outside France, continued to fight for French freedom. Cousteau was working for Naval Intelligence against the occupying powers. To cover his activities, he managed to obtain a permit from the occupation authorities enabling him to shoot cultural films in the Mediterranean. But his chief object was to develop his ideas for a self-contained underwater breathing apparatus.

In December, 1942, Cousteau met in Paris with Émile Gagnan, an expert on industrial gas equipment. When Cousteau had finished explaining his needs to the engineer, Gagnan handed him a small mechanism made of plastic. "Something like this?" he asked, "It is a demand valve that I have been working on to feed cooking gas

Above left: a diver using a snorkel examines coral in shallow water off the Kenya coast. He can move freely but must return to the surface every two or three minutes for a gulp of air.
Above right: a diver supplied with air from the surface. He can stay down for a long time but can only move within the limits imposed by his life line.

automatically into the motors of automobiles." Gasoline was in short supply in wartime France and many experiments were going on to find a substitute for it.

Within a few weeks Cousteau and Gagnan had completed work on their first demand valve, or regulator, for human breathing. Cousteau tried it out in the wintry waters of the Marne River near Paris. He was able to breathe easily, even though most of the air bubbled wastefully from the regulator. Then Cousteau tried standing on his head. The air supply slowed almost to a halt. Cousteau could not breathe. The first attempt to free man in the sea was a failure.

But in the time it took to drive back to Paris, the two men hit upon a solution to the problem. The exhaust and intake outlets in the experimental regulator were six inches apart. When Cousteau was standing or swimming horizontally, the exhaust was above or at the same level as the air inlet. But when he swam vertically downward, the exhaust was six inches lower, and the tiny difference in pressure blocked the inflow of air. The solution was simple. Place the exhaust as close to the inlet as possible to minimize the pressure difference. It worked—at least in a test tank in Paris. But would the regulator function in the sea at the depths and pressures that Cousteau hoped to penetrate? Only a real dive into the Mediterranean would provide the answer.

On a June morning in 1943, Cousteau went to the railroad station in the tiny town of Bandol on the French Riviera to pick up a wooden case shipped from Paris. Nearby was a large seaside villa that he had rented as a base for diving operations. Cousteau had

invited his diving companions, Philippe Tailliez and Frédéric Dumas, to join him there. The group had chosen the most obscure site they could find for the first test dive with Cousteau's new apparatus. Germany and Italy were using military frogmen in the war. Their swimming soldiers were equipped with an early type of oxygen lung that was dangerous to use and limited their diving range to 40 feet. The German Admiralty was spending millions of marks on experiments to develop more advanced underwater equipment. If Cousteau's new device worked it would be far superior to any underwater lungs the Germans had so far devised.

As soon as the box arrived, Cousteau rushed it to the Villa Barry where Tailliez and Dumas were waiting. "No children ever opened a Christmas present with more excitement than we did when we unpacked the first 'Aqua-Lung'," says Cousteau, describing the moment in his book *The Silent World*. "We found an assembly of three moderate-sized cylinders of compressed air linked to an air-regulator the size of an alarm clock. From the regulator there extended two tubes, joining on to a mouthpiece. With this equipment harnessed to the back, a watertight glass mask over the eyes and nose, and rubber foot fins, we intended to make unencumbered flights in the depths of the sea."

Cousteau, as head of the group, was to be the first to test the new Aqua-Lung. "Didi" Dumas, whom Cousteau called the best goggle diver in France, was to wait on the beach, ready for an instant rescue if anything went wrong. Cousteau's wife, Simone, was to swim on the surface above him, watching the experiment through her mask.

Above: a diver wearing an Aqua-Lung. He can move freely like a snorkel diver and yet can stay down for hours at a time. The cylinders of compressed air on his back are connected to his mouth-piece by way of a vital piece of equipment—a demand regulator. As the diver breathes, this feeds him the exact amount of air he needs.

Dumas and the others strapped the cylinders onto Cousteau's back and attached seven pounds of lead ballast to his belt. Cousteau waded through the surf "with a Charlie Chaplin waddle" and then sank gently into the silent waters of the Mediterranean.

"I breathed sweet, effortless air," wrote Cousteau. "There was a faint whistle when I inhaled and a light rippling sound of bubbles when I breathed out. The regulator was adjusting pressure precisely to my needs.

". . . A modest canyon opened below, full of dark green weeds, black sea urchins and small flowerlike white algae. . . . The sand sloped down to a clear blue infinity. . . . My arms hanging at my sides, I kicked the fins languidly and traveled down, gaining speed, watching the beach reeling past. . . . I reached the bottom in a state of excitement. . . . I looked up and saw the surface shining like a defective mirror. In the center of the looking glass was the trim silhouette of Simone, reduced to a doll. I waved. The doll waved at me. . . .

"I experimented with all possible maneuvers of the Aqua-Lung—loops, somersaults, and barrel rolls. I stood upside down on one finger and burst out laughing—a shrill distorted laugh. Nothing I did altered the automatic rhythm of air. Delivered from gravity and buoyancy, I flew around in space. . . . I went down to 60 feet. We had been there many times without breathing aids but we did not

Above: Jacques-Yves Cousteau receiving an award for his underwater film *Le Monde Sans Soleil* (sunless world), made in 1964. This is one of the films Cousteau has made of his explorations of the world under the waves.

Right: a diver examines the hull of an old ship. In 1943, Cousteau realized how valuable his diving equipment would be in underwater salvage work. He made a film which he called *Épaves* (sunken ships), photographing underwater wrecks off the coast of southern France. While making the film, he and his team pushed their maximum diving depth beyond the 130-foot mark.

know what happened below that boundary. How far can we go with this strange device?" In the months and years to follow, Cousteau and other divers were to find out.

During that first exciting summer, Cousteau and his team made 500 successful dives to depths between 50 and 100 feet. The Aqua-Lung behaved perfectly and the men had a whole new world to discover. Yet Cousteau still felt apprehensive. "The thing was too easy," he wrote. "Every instinct insisted that we could not so flippantly invade the sea. An unforeseen trap awaited in the deep, any day now, for Dumas, Tailliez, or me."

It was still wartime. The previous year, German troops had invaded the naval base at Toulon, the Mediterranean port in southeast France. The French fleet had destroyed most of its ships in the harbor. The scuttled vessels included two of those on which Cousteau had served. "Sunken ships preyed on our minds," said Cousteau. In 1943, he and his team began to make a film about undersea wrecks. The film was also to serve as proof of the usefulness of the Aqua-Lung in underwater salvage work.

Among the wrecks explored by the divers was the *Dalton*. This British steamer was under charter to a Greek company when she sank off Marseille on Christmas Eve, 1928. Part of the ship lay from 70 to 100 feet down, the stern quarters were about 30 feet deeper. The divers hesitated, wondering whether to take the risk of going

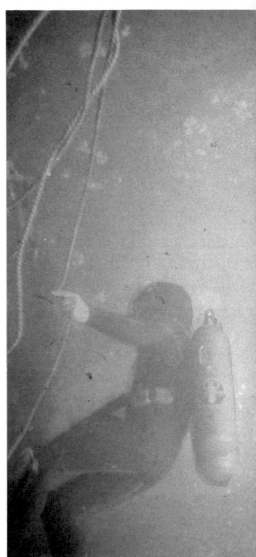

Above: a diver pulling ropes down from the surface during work on wrecks off the coast of England.

Below: a peril for the skin diver, a turkeyfish *(Pterois volitans)*. Its dorsal spines are poisonous.

Above: a diver's back pack opened up to show the cylinders. They contain a mixture of helium and oxygen. This mixture is used for dives deeper than 150 feet because helium under pressure does not affect the nerve cells as does nitrogen—the major constituent of compressed air. Cousteau and his team experienced the narcotic effect of compressed nitrogen. They called this effect "the rapture of the depths."

deep enough to investigate the tantalizing stern section. Cousteau decided that this was the only way to ascertain the limits of the Aqua-Lung. Down they went, deeper than ever before. They reached 132 feet and returned safely.

But Dumas, in particular, believed that the Aqua-Lung could take men deeper still without risk from the crippling effects of the bends. He was ready to test his belief by going down to find out.

On October 17, 1943, preparations were made for an experimental dive. A rope, knotted at regular intervals, was carefully measured and attested by a local official. The rope was lowered to the seabed, about 240 feet down, at the spot where Dumas would attempt a record-breaking dive. Two launches, full of witnesses, accompanied the "condemned man," as Cousteau called him. Dumas was to submerge, heavily weighted, until he reached his limit. Then he was to fix his weights to the line and rise to the surface as quickly as possible.

Dumas dived into a cold and choppy sea. Cousteau, the safety man, dropped to his vantage post 100 feet down on the line, after a breathless struggle against the waves. He wrote, "I followed him [down]. . . . My brain was reeling. Didi did not look up. I saw his fists and head melting into the dun water."

Dumas described his record dive in these words: ". . . . I cannot see clearly. Either the sun is going down quickly or my eyes are weak. I reach the hundred foot knot. My body doesn't feel weak but I keep panting. The damned rope doesn't hang straight. It slants off into the yellow soup . . . . I am anxious about that line, but I really feel wonderful. I have a queer feeling of beatitude. I am drunk and carefree. My ears buzz and my mouth tastes bitter. The current staggers me as though I had had too many drinks.

"I have forgotten Jacques and the people in the boats. My eyes are tired. I lower myself farther, trying to think about the bottom, but I can't. I am going to sleep, but I can't fall asleep in such dizziness. There is a little light around me. I reach for the next knot and miss it. I reach again and tie my belt on it.

"Coming up is merry as a bubble. . . . The drunken sensation vanishes. I am sober and infuriated to have missed my goal. I pass Jacques and hurry on up. I am told I was down seven minutes."

Dumas thought that he had gone no more than 100 feet down. But his weighted belt was tied off 210 feet below the surface. The

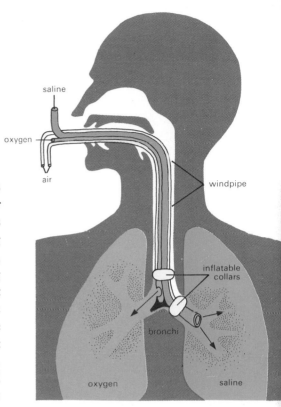

Right: the diagram illustrates an experiment performed by Johannes A. Kylstra. It was carried out in the light of Cousteau's suggestion that a type of underwater man could be created, who would breathe in the sea as easily as a fish. Kylstra filled one of a deep-sea diver's lungs with a saline solution that was "breathed" by being pumped in and out. The diver later said that he felt no unpleasant sensations.

saline

oxygen

air

windpipe

inflatable collars

bronchi

oxygen

saline

Frenchman had discovered another danger of the sea, one as perilous as the bends. Cousteau's team called it *l'ivresse des grandes profondeurs*—the intoxication, or rapture, of the depths. This "drunkenness" becomes progressively worse with depth. A feeling of elation is succeeded by drowsiness and overwhelming lethargy, while the diver's thoughts become gradually more confused. Feeling perfectly safe, divers have been known to tear off their equipment while "drunk" on nitrogen. Some divers are even reported to have offered their breathing apparatus to a passing fish, lest the creature should drown without air. Others have searched in non-existent pockets to make sure they have brought their cigarettes with them.

This "drunkenness," or nitrogen narcosis, is apparently caused by the effect of nitrogen on the nervous system. Today, deep divers

Left: a close-up of a modern diver's helmet. Compared to the heavy, uncomfortable diving helmets of the 1800's this lightweight version is very comfortable to wear. Good visibility is provided by a large, wrap-around visor. Tubes from the gas cylinders enter the helmet from the sides and are connected to a standard mouthpiece.

use a mixture of helium and oxygen rather than nitrogen and oxygen to guard against the rapture of the depths. But in 1943 the depths of the sea were all too new. Each lesson had to be personally learned.

When the war ended, Cousteau showed his film of sunken ships to the Ministry of the Marine in Paris. As a result, he was given the task of organizing an underwater group to salvage vessels sunk during the war and to carry out diving research. With Tailliez and Dumas, Cousteau set up the Undersea Study and Research Group based on Toulon, where other naval officers and technical experts joined them. Their work was made doubly hazardous by the number of unexploded mines which littered the seabed. Dumas, curious about the effects of explosions underwater, came near to death on more than one occasion during reconnaissance dives.

The group's activities gradually expanded. Other tough assignments came from the Navy. Cousteau and his men continued to dive, film, and experiment, perfecting their techniques as they worked. But the rapture of the depths still fascinated them. In the summer of 1947, the Undersea Research Group decided to challenge the sea

again in a series of even deeper dives. The depth they set for themselves was 300 feet. In the four years since Dumas had dived to 210 feet, no independent diver had beaten his record.

Cousteau was the first to dive. Grasping heavy scrap iron, he plunged down through the water, watching the azure blue of the sunlit surface dim as he neared the lightless depths. At 200 feet, Cousteau "tasted the metallic flavor of compressed nitrogen and was instantaneously and severely struck with rapture." Grabbing the line, Cousteau jotted on one of the small message boards strung along the rope, "Nitrogen has a dirty taste."

"I hung witless on the rope," he wrote later. "Standing aside was a smiling jaunty man, my second self, perfectly self-contained, grinning sardonically at the wretched diver. As the seconds passed, the jaunty man installed himself in my command and ordered that I unloose the rope and go down." Cousteau dropped to the last board, 297 feet down. He signed his name, dropped his load of iron ballast, and shot upward. He had become the deepest free diver in the world.

Then, five more of the group divers visited the deep board. Not

Above: the two center divers are completing tests during Cousteau's first underwater living experiment. Cousteau was fascinated by the idea of men living in the sea. In 1962, he sent down two divers to live for a week in a habitat 33 feet below the surface. The experiment was a complete success and Cousteau followed it up with a series of more ambitious missions.

Left: a diver is helped back into a rubber dinghy. He is one of several divers who surveyed wrecks off England's southwest coast in 1967.

Right: a diver using an underwater scooter. These devices not only provide an exhilarating ride but they also save precious seconds of the limited time a diver has underwater.

one of them was able to scrawl anything intelligible at that depth.

The summer passed and the team decided on another series of dives past the 300-foot mark. The first to go down was Maurice Fargues, a diving master in the group. Down he plunged, occasionally tugging the rope in an "all's well" signal. Suddenly, there was nothing. The safety man dived immediately to meet him at the 150-foot level while others hauled Fargues toward the surface. When the two divers met, Fargues' mouthpiece was dangling on his chest. He was dead. Later, the group examined the boards on the rope and found Fargues' initials scratched at a depth of 396 feet. It was 100 feet deeper than any of the others had ever gone. The group never again challenged the rapture of the depths.

In 1949, Cousteau took command of the *Calypso*. This vessel was designed to undertake a variety of oceanographic research. In 1952, Cousteau took it to Grand Congloué, an islet off Marseille, where he and his team explored one of the most interesting wrecks ever discovered—a 100-ton freighter dating from about 230 B.C.

It was while working on this wreck that Cousteau developed the idea of underwater television. The archaeologists on board the

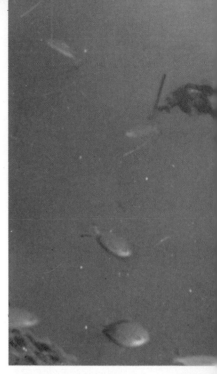

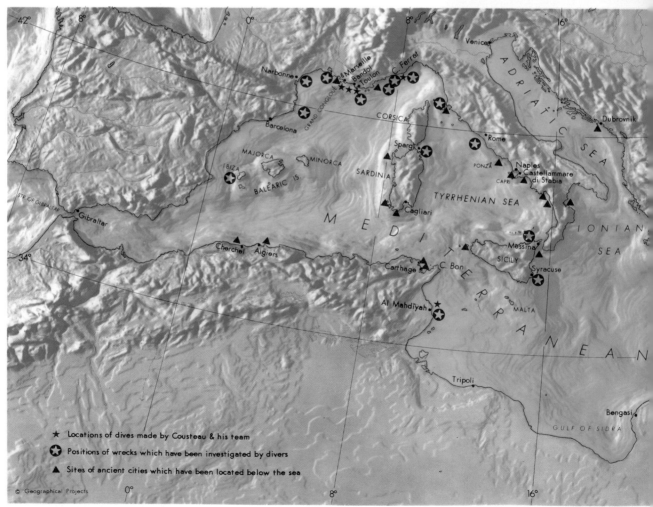

★ Locations of dives made by Cousteau & his team

✪ Positions of wrecks which have been investigated by divers

▲ Sites of ancient cities which have been located below the sea

© Geographical Projects

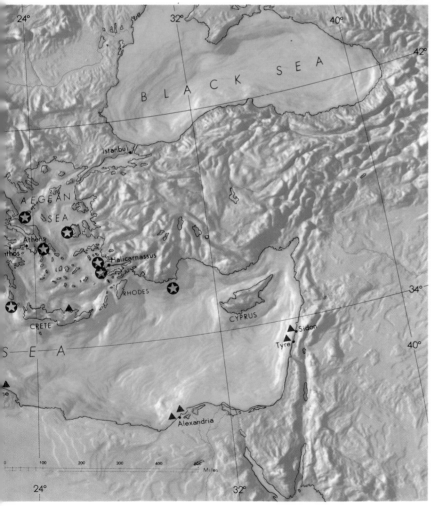

Above: Cousteau's diving saucer. This free-swimming submersible is propelled by water jets. It can carry two men underwater to a depth of 1,000 feet. Below: an Aqua-Lung diver scoops up a vase from the bottom of the sea.

Left: the Mediterranean Sea, showing underwater exploration. At Bandol in 1943 Cousteau and his team tested the first Aqua-Lung, an invention that was to revolutionize deep-sea diving. For the first time, man was able to dive beneath the surface of the waves unencumbered by airlines linking him to the surface.

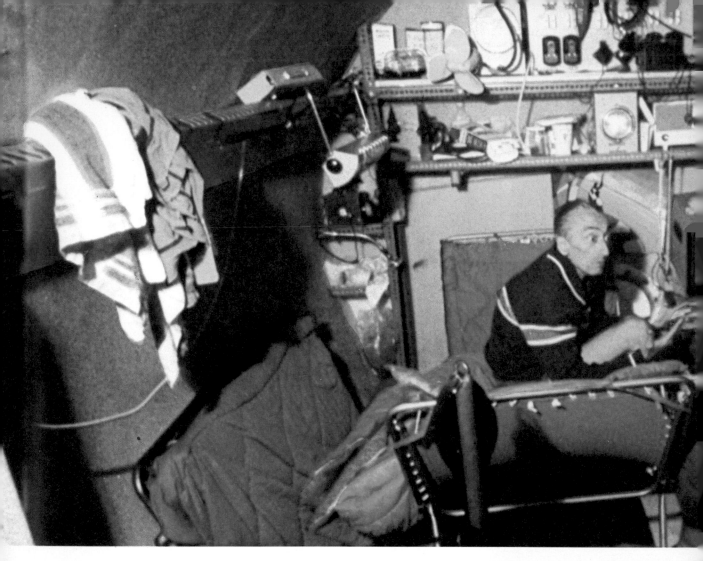

Calypso were anxious to participate more directly in the work of the divers below. Cousteau arranged for the divers to take down a tele-camera, which he borrowed from a large television company. The camera was fitted with a watertight case and could be operated either by the diver or from the surface. Underwater microphones carried instructions down from the *Calypso,* where spectators received a clear picture on their television screen of the activities below.

This was just one of the many devices that Cousteau developed over the years to improve and advance underwater research. Notable among the others is his diving saucer, a two-man submarine. The saucer is equipped with outside mechanical arms and is propelled by water jets.

In the thousands of hours that Cousteau and his men had spent diving under the sea, they had really no more than glimpsed its fascinations. They were men-fish for only a few hours at a time. In a submersible, they could stay longer and dive deeper. Yet, then they would no longer be *in* the sea, but confined within a steel prison. To set up a "home" on the sea floor that would let men live and work

Left: Cousteau inside his first underwater habitat *Conshelf I.* In September
1962, two divers, Albert Falco and
Claude Wesley, stayed in *Conshelf I*
for a week. The habitat was at a
depth of 33 feet and the divers used
it as a home base for their daily
excursions into the surrounding sea.

Below: a view of the outside of
Conshelf I. The 17-foot-long cylinder
was anchored to the seabed with 34
tons of pig iron. Divers entered and
left the habitat through an open hatchway in the bottom of the cylinder.
The *Conshelf I* experiment showed that
men could live and work underwater.

in comfort there would be the next step in making man into a truly marine creature.

Cousteau **was** a pioneer in developing such a manned underwater station in the sea. In September, 1962, he set up the world's first underwater habitat. It was called *Conshelf I,* short for Continental Shelf Station Number One. Two aquanauts from his team lived and worked for a week in the habitat, anchored 33 feet underwater off the coast of southern France.

Conshelf I was a steel cylinder 17 feet long and 8 feet in diameter. The entrance to the cylinder was an open hatchway in the floor of the habitat. No door was needed because the air pressure within the cylinder was exactly that of the sea outside. The aquanauts worked for five hours each day to show that men could not only live under-water but also perform useful tasks.

As Cousteau wrote about *Conshelf I*: "They [the divers] were supported—or rather over-supported—by a clutter of vessels and men; as other divers, and just plain well-wishers dropped in, they sometimes felt as if they were living in a bus station." Nevertheless, the divers were reluctant to emerge after their week in the sea, and were in excellent health. The first step had been taken toward living in the sea.

In the same week that Cousteau was carrying out his experiments

Below: *Conshelf II* on the remote Sha'ab Rumi Reef in the Red Sea. Cousteau's second underwater habitat consisted of a starfish-like complex of cylinders that looked more like a village than a house. In 1963, five men stayed in the habitat for a month. Cousteau was not so much interested in breaking depth records—*Conshelf II* was only 36 feet below the surface—as in setting up the forerunner of a possible future "underwater society."

with *Conshelf I,* a Belgian diver, Robert Sténuit, was trying out another underwater home. This was a 3-foot by 11-foot aluminum cylinder designed by American inventor Edwin Link. Sténuit remained for 26 hours at a depth of 200 feet, spending some of that time exploring the sea outside. After his successful stay in the sea, the American program of underwater habitats forged ahead.

Six months later, Cousteau followed his first success with *Conshelf II,* a "village" of five interconnected cylinders and an "advance station" of a single cylinder. His aim this time was to set up a "precursor of underwater society." The main settlement of this colony, 36 feet beneath the sea, housed five men for a month. In the deeper camp at 90 feet two men lived and worked for a week, pushing the depths at which men could do useful work down to 165 feet.

Conshelf II was remarkable for another finding. Cousteau had assembled a group of "average" men for his experiment. He proved that the participants did not have to be expert divers or even young men in top physical condition. Instead, the men of *Conshelf II* were chosen for their individual skills, either as mechanics, cooks, or scientists. And on this second experiment in underwater living, Cousteau took care that the divers were not bothered by underwater sight-seers.

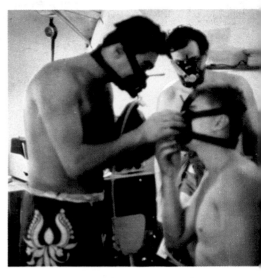

Above: divers fixing each other's masks before swimming out from their underwater home *Conshelf III.* At a depth of 330 feet in the sea their preparations had to be very thorough. If they got into difficulties they had to be able to return to the habitat. To swim to the surface from depth without decompressing would be fatal.

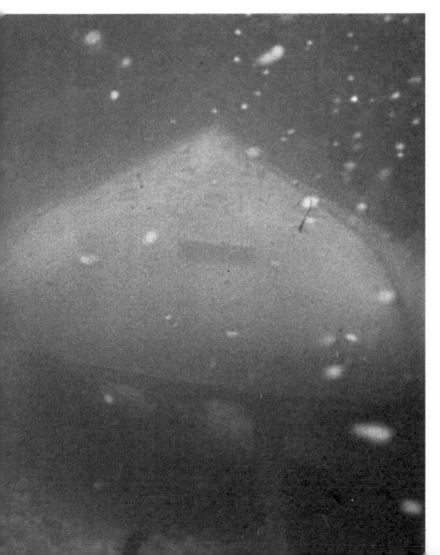

Conshelf III, in 1965, was the deepest and most ambitious of Cousteau's experiments toward the ultimate dream of *Homo aquaticus,* or underwater man. It was to be the most advanced outpost yet established by man in the "offshore wilderness," an attempt to open up hundreds of thousands of square miles of the planet's surface for the use of man. The new habitat was to be located 330 feet down and to house six men, or oceanauts, as Cousteau called them. The plan called for them to live and work in the darkness for more than three weeks.

While conceiving *Conshelf III*, Cousteau decided that the real risk to men living underwater came from the cables and pipes that provided an "umbilical cord" to the surface above. These slender cables providing air to breathe were vulnerable to surface storm damage or some other accident. Better that the men of *Conshelf III* should live in a completely self-contained environment with virtually no reliance on those at the surface except in dire emergency.

The new undersea station was a sphere 18 feet in diameter resting on a 28-foot chassis that held 77 tons of ballast. Before being towed to their destination off Cape Ferrat in the Mediterranean, the six highly-trained oceanauts were pressurized in *heliox*—a helium and oxygen mixture—to a pressure of 11 atmospheres. While still on the surface, they were living under the equivalent of 330 feet of water. But although the heliox mixture protects against the bends and against the rapture of the depths, it poses another problem—that of communication. Helium is so light that it does not "slow down" the vibrations of the vocal cords as does normal air. The speech of people breathing a helium mixture sounds, as one ocean scientist put it, "like Donald Duck in a rage." Cousteau's oceanauts were almost completely unable to understand one another for days.

Cousteau considered that there were two critical periods for *Conshelf III*, apart from unexpected emergencies. One was the descent and precise landing of the huge sphere at a preselected site on the bottom. The other was the return of the habitat to the surface under the control of the oceanauts themselves.

The first critical phase passed successfully. The habitat landed within a foot of its target site. Now came the difficult routine of work with a five-ton mock-up of an oil wellhead, the elaborate system of valves and piping that controls the flow of oil from an underwater well. Conventional divers cannot maintain a wellhead,

Above: *Conshelf III.* Cousteau's third habitat consisted of an 18-foot-diameter sphere divided into two floors, the upper one for eating and scientific work, the lower one for sleeping and diving. Six oceanauts (including Cousteau's son Philippe) lived in *Conshelf III* for three weeks at a depth of 330 feet below the surface.

Right: *Conshelf III* divers working on a mock-up of an underwater oil well-head. One of their main tasks was to discover if divers could maintain these structures at depths below 150 feet.

or *Christmas tree* as it is called, in depths much below 150 feet. For the *Conshelf III* oceanauts to succeed in the difficult job of deepwater maintenance would prove that oil could be produced deeper than ever before—and that such tasks as underwater mining, salvage, and aquaculture (farming the sea) could be done as well.

For three weeks, the six divers went through their daily routine, working with the Christmas tree, observing marine life, and venturing from their base as deep as 370 feet. But then it was time to return. The next critical period was at hand.

Sealed in the sphere, the chief diver turned a crank to release ballast that would free the sphere from the bottom and send it slowly drifting upward. As the iron weights were released, they turned the sea floor into a maelstrom of swirling sediment. But the sphere did not move. The undersea habitat was stuck fast to the bottom.

Cousteau, connected to the habitat by telephone from the command station ashore, suggested that a little compressed air be squirted into the ballast tanks, just enough to loosen the habitat but not enough to send her careering upward. André Laban, commander of the sphere, cracked the compressed air valve for two seconds. "She's not moving," he reported. Cousteau suggested "another gentle injection." There was still no result.

"Here goes a little more," said Laban, twisting the valve again. *Conshelf III* trembled. Then, gently, the sphere began to rise. The habitat was free at last. Minutes later, the huge sphere bobbed on

Above: recent research has shown that glass is as strong as steel in its power to resist the water pressure at great depths in the ocean. But molding and joining pieces of glass to make flawless spheres has always presented problems. The glass sphere shown here rests in a protective cage while being prepared for pressure chamber tests. In use it may be lowered to depths of up to 1,500 feet or may form part of a self-powered submersible. Right: glass is also to be used in the deep-sea city that General Electric hope to start constructing in 1980. Unlike steel, glass does not corrode in seawater, an important consideration in permanent underwater habitats. The drawing shows how standard 12-foot-diameter glass spheres fit together to make a complete self-contained city—12,000 feet down in the Atlantic Ocean. Divers wishing to swim around their home will carry devices in their lungs so that they can "breathe" water.

the surface of the ocean. The divers inside would have to undergo 84 hours of decompression to adjust their bodies slowly to normal air pressure. But they—and the visionary Cousteau—had scored an incredible success.

The *Conshelf* experiments continue today, pursuing ever greater and more ambitious objectives. Each one will mark another milestone on the road to the deep. And each will be a further tribute to the dedicated pioneering of the first aquanaut—Jacques-Yves Cousteau.

Above: the standard sphere—glass segments "seamed" with titanium.

A New History of the Earth

8

Free divers and submersibles are adding every day to man's knowledge of the undersea world. The ability to "go down and see" offers seemingly boundless possibilities for the exploration of the deep. But a great deal of oceanographic research is still carried out from the surface. Ships are playing a vital role in discovering fresh facts about the sea and, through them, gaining a greater understanding of our planet.

One such ship is the *Vema,* possibly the most productive oceanographic vessel in operation and with a long history of valuable research behind it. For the *Vema* has worked in every ocean of the world, logging more miles than any other oceanographic ship of its size.

A 200-foot, 3-masted steel schooner, the *Vema* was built in 1927 as a rich man's yacht. Its varied career later included trade runs between Canada and the West Indies, taking out cargoes of lumber and returning with stocks of West Indian rum. In 1953 the *Vema* was purchased by Columbia University's Lamont-Doherty Geological Observatory and fitted out to serve the cause of oceanography. It was refurbished with a steel deck and deck housings, but is still without proper lifeboats (its Canadian crews claim that dories—narrow, flat-bottomed boats—are far better at sea) or any provision for distilling fresh water. It does not even have watertight bulkheads. Some oceanographers have called the *Vema* and its cramped quarters a floating coffin. Many others think it the most seaworthy ship afloat.

The *Vema*'s early years as a floating laboratory were stormy ones. For months on end it shuddered under the pounding of high explosives that sent showers of water crashing against its steel hull. The oceanographers aboard it became oblivious to the sound. But on another ship, many miles away, the echoes of the explosives were being carefully recorded. After a time the ships would exchange roles—the *Vema* would listen while explosives were set off from the other ship. The scientists on the two ships were engaged on a mission to probe the sea floor with its thick layers of sediment and

Left: an island is born as an erupting volcano breaks surface 150 miles off the Pacific coast of Japan. Scientists know that the earth's crust is thin under the oceans, but only recently have they been able to probe its secrets.

Above: geophysicists examine a profile of the sea bottom produced by seismic soundings. The trace shows underwater mountains called seamounts.

to penetrate the solid rocks below, in search of clues to the earth's past—buried millions, perhaps billions, of years before. With high explosives and film, they were trying to wrest the history of the planet from the depths of the sea.

A charge of high explosives has a tremendously powerful impact in water, far greater than in the comparatively thin air. Energy from the explosion travels outward in all directions. Some of it moves toward the bottom at the speed of sound in water—usually about 4,700 feet a second or around four times faster than in air. When the energy from the explosion reaches the bottom, most of it bounces off. But some of the energy knifes through the bottom until it reaches a layer of different matter beneath the sea floor and part of it bounces off again. Even then the remaining energy manages to penetrate the layer and the successive layers that lie beneath it. At last, the sound may echo off the *basement* itself, the solid rocky bottom of the ocean formed before any material had sifted through the water to cover it.

As the sound energy hits these different layers beneath the sea floor, the sound not only bounces off but also travels along the layer, radiating noise like a plane traveling overhead. This sound broadcasting from the different layers can be picked up by a listening ship equipped with sensitive *hydrophones*—microphones that work in the water. By noting exactly the time it takes for the different echoes to reach the hydrophones of the listening ship, marine geophysicists can tell how deep these layers are, whether they slant up or down or are level, and even make a good guess at what the layers are made of. This technique of reading the layers under the sea bottom requires absolute teamwork between two ships connected only by radio.

The shooting ship, as the ship which drops the explosive charges is called, must put the charges over the side on a precise schedule. The listening ship must know almost to the split second when the charge will go off. This is because any sound—the rumble of a generator, the sound of a pump, even the clank of a wrench dropped on deck—can be picked up by the hydrophones. And this unwanted noise can obscure a vital trace on the film that records the seismic signals—signals caused by an earthquake or an artificial vibration of the earth, in this case the explosion. Absolute silence is the rule on the listening ship.

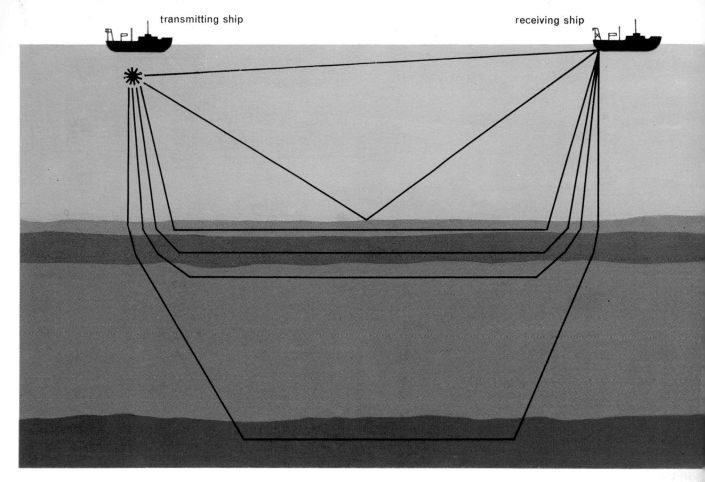

transmitting ship   receiving ship

The same sense of urgency exists on the shooting ship. Handling high explosives at sea is dangerous and the shooting schedule is fast. When the shooting ship is near the listening ship, half-pound blocks of explosives are tossed over the side every 30 seconds or so. To keep up this pace, the man setting off the explosives must cut a six-inch length of fuse, crimp an explosive cap to one end, tape the cap to the charge, and ignite the fuse—twice each minute. But as the ships separate, the rate of shooting drops as the size of the individual charges becomes larger. One-pound charges follow half-pounders, then 3-pounders, 9-pounders, 24-pounders, 100-pounders, and finally, at 20-minute intervals, large depth charges loaded with several hundred pounds of high explosive. When even the depth charges are too weak for the bottom echoes to be heard, the shooting ship heaves to and becomes the listener, the ships move closer together and the series of explosions begins again.

Above: the seismic-refraction method of probing the layers of sediment beneath the sea. One ship releases an explosive charge in the water and another, up to 100 miles away, records the refracted sound waves and measures how long they have taken to arrive. Scientists use the traces produced to tell the depth of the layers beneath the seabed, whether they slant up or down, or are level, and what they are made of.

Below: a section through the floor of the Pacific Ocean. Nicaragua and the Philippine and Marshall island groups are the peaks of undersea mountains. Challenger Deep, at the left, is the deepest known spot on earth.

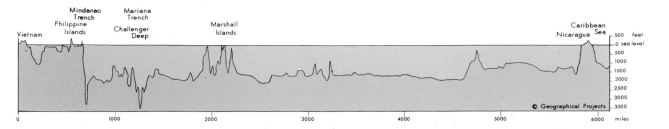

Above: oceanographers of the British
research ship *Discovery II* maneuver a
sonar "fish" into the water. As it
is towed behind the ship the fish
"illuminates" the sea floor with sound
waves and picks up the returning
echoes. On board *Discovery* a display
unit translates these echoes into an
acoustic "picture" of the sea floor,
showing features many thousands of
feet below. Using such devices
oceanographers have been able to
make accurate surveys of the ocean
bottom in a fraction of the time taken by
more conventional sounding methods.

The *Vema* worked in this way with dozens of other ships, regis-
tering thousands of miles of seismic tracks. But in recent years,
techniques aboard it have changed. The hazardous high explosives,
which caused the death of one oceanographer, are no longer used.
Today, an air gun slung from the stern is used instead. The long
strands of film used to record the signals have been replaced by an
electronic instrument that automatically sketches a picture of the
layers below the sea floor.

One piece of the *Vema*'s equipment that has not become obsolete
is the Ewing piston corer, named after its designer Maurice Ewing,
the director of the Lamont-Doherty Geological Observatory of
Columbia University. Ewing, who has sailed regularly aboard the
Vema since 1953, was the first to use seismic techniques at sea.

The Ewing corer is used for obtaining sediment samples. It
consists of a hollow tube that is driven into the bottom sediments
by a 2,000-pound weight. Inside the tube is a movable piston that
sucks the sample into the tube without disturbing the sediments.
The column of layered sediments that is pushed from the tube onto
the deck of the *Vema* contains clues to the geological history of the
bottom and to the dynamic events taking place on the ocean floor.
Some of the many thousands of cores at Lamont are longer than
70 feet and contain material laid down as long as 100 million years
ago.

Before Ewing perfected his corer, other marine scientists had
tried punching holes in the deep ocean bottom. Their short cores
revealed little more than the *Challenger* had discovered in the 1800's.
The bottom of the deep sea was thought to be a place of perpetual
calm with a constant rain of sediment slowly blanketing it. But
Ewing's cores revealed that the ocean floor was far less peaceful
than earlier oceanographers had believed. Working with David
Ericson of Lamont, Ewing discovered that vast areas of the deep
ocean basins had been covered again and again by enormous
undersea landslides that swept down from the shelves ringing the
continents or from seamounts poking upward from the bottom.
Other scientists found that these slides must have rushed across the
sea floor at speeds of more than 100 miles an hour, smothering the
marine life on and just below the surface of the bottom ooze.

Another veteran instrument aboard the *Vema* is the PDR—the
Precision Depth Recorder—which traces a constant graphic outline

PHYSIOGRAPHIC DIAGRAM OF THE
SOUTH ATLANTIC OCEAN
The Caribbean Sea, The Scotia Sea, and the eastern margin of the South Pacific Ocean

BY BRUCE C. HEEZEN AND MARIE THARP

LAMONT GEOLOGICAL OBSERVATORY
Columbia University

Left: a view from the derrick of the drilling ship *Glomar Challenger.* Thousands of feet of drill pipe are laid out in precise order on deck.

Right: the 400-foot-long *Glomar* on station over a drilling site. Named for the original *Challenger,* this remarkable ship has enabled ocean-ographers to make astounding discoveries about the history of the earth.

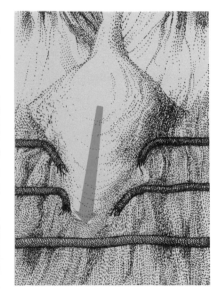

Above: a diagram showing how an underwater landslide snapped a series of telegraph cables off Newfoundland in 1929. By measuring the time interval between the break in each of these cables, oceanographers worked out that the slide was moving at a maximum speed of 50 miles an hour.

of the shape of the sea floor. It was this PDR that helped oceanographers discover one of the strangest features of the undersea world—the vast mid-ocean ridge that forms the greatest mountain range on earth.

As early as the *Challenger* expedition, oceanographers recognized that there was a peculiar rise in the middle of the Atlantic Ocean. As more and more soundings were taken, this hump in the Atlantic gradually took on the shape of an extended mountain chain. Most of the top of the ridge was found to be between 9,000 and 10,000 feet deep, some 5,000 feet above the average depth of the Atlantic Ocean, but in places its peaks rise above the surface of the water to become islands such as the Azores in the North Atlantic, St. Paul Rocks in the South Atlantic about 600 miles northeast of Brazil, and Ascension, Tristan da Cunha, and Bouvet, also in the South Atlantic. The tallest of these peaks is Pico Island in the Azores which is 27,000 feet high, with 7,460 feet above the surface.

Although the main outline of this Atlantic ridge was known by about 1930, no one knew where it began or where it ended. Later, other ridge systems were found in other oceans. But no one thought to connect them into a single giant chain. Then Ewing suggested that all the ridges were part of one enormous "seam around the world" that was of great importance in understanding the deepest secrets of the earth itself. Just how, he was uncertain. But he was sure that the mid-ocean ridge was a vital key to our past. One part of this chain, however, was missing—a ridge to join the peaks of the Atlantic with those in the Indian Ocean. The *Vema* and its PDR were assigned the task of finding this missing link.

The *Vema* picked up the known part of the ridge in the South Atlantic and headed southeast toward the border of the Atlantic and Indian Oceans, south of the Cape of Good Hope. Watch by watch, scientists peering at the PDR saw the missing sections of the mid-ocean ridge gradually unreeling. The *Vema* sailed into Cape

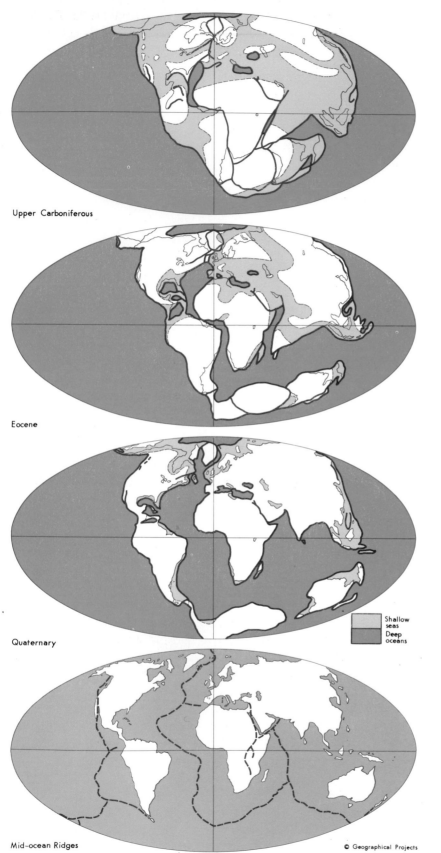

Upper Carboniferous

Eocene

Quaternary

Shallow seas

Deep oceans

Mid-ocean Ridges

© Geographical Projects

Left: according to certain scientists the continents of the world were once joined together in one huge land mass, but have since drifted apart to their positions today. This theory is known as *Continental Drift*. These maps show how the continents may have drifted apart. From top to bottom: the position of the continents during the Upper Carboniferous Period (310 million to 275 million years ago); during the Eocene Epoch (55 million to 40 million years ago); and during the Quaternary Period ($3\frac{1}{2}$ million years ago to the present day). The bottom map shows the position of the mid-ocean ridges, an undersea mountain chain forming a seam around the earth.

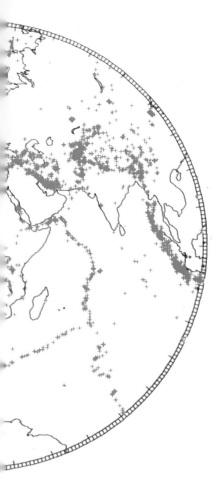

Above: a computer-drawn globe showing the location of earthquakes over a period of several years. The pattern of dots agrees with the theory that the outer crust of the earth is divided into a number of plates that are constantly moving. Accordingly the earthquakes occur where disruption is greatest, at the edges of the plates.

Town, South Africa, for supplies, and then out again, still following the elusive ridge south of Africa around toward the Indian Ocean.

At last, the underwater heights were proved to join others that had already been charted. A few small gaps still existed. But the *Vema* had shown that the globe-girdling chain was a reality. Soon another ship was to show how important the ridge system is to our knowledge of the earth. This vessel, with Maurice Ewing and many other distinguished oceanographers on board, has also furthered the *Vema*'s pioneering studies of the sea floor, revealing the nature of the mysterious layers that lie beneath it. Its work in the deep oceans has made it one of the most important research vessels afloat today.

This remarkable ship is the *Glomar Challenger,* an ungainly-looking, 400-foot-long drilling vessel, topped by a tower of steel. Built in 1968, it was designed expressly to probe through miles of water into the "pages" of the bottom sediments that are the earth's archives, and deeper still into the original floor of the sea. The ship is a worthy descendant of that earlier *Challenger,* which uncovered so many of the secrets of the sea. Manning the *Glomar* is a new breed of seafarer—a rugged crew of deep-ocean drillers who have perfected their skills on demanding and dangerous offshore oil rigs. Teamed with the drillers are seasoned sailor-scientists who have studied the oceans for years. Together, scientists, ship, and crew have penetrated the deep sea floor to answer some of the most fundamental and puzzling questions about our planet.

Ever since the first maps of the world were made, men have been struck by the apparent jigsaw fit of the continents. Africa nestles neatly into the curve of South America. Greenland fits snugly into the top of the joint between North America and Europe. If Antarctica were joined to South America, Australia would fit neatly against its western coast, and India would complete the jigsaw between Africa and Australia.

It was not until 1912, however, that the idea of *Continental Drift* was put forward by Alfred Wegener, an Austrian scientist. Wegener compared the continents to icebergs drifting on a sea of soft rock. He claimed that Australia, Africa, India, South America, and Antarctica were once joined together in a supercontinent which he called Gondwanaland. Eurasia, Greenland, and North America were similarly connected together in another giant continent that Wegener named Laurasia. Earlier still, Wegener suggested, these two supercontinents were joined as a single land mass called Pangaea.

But Wegener was unable to explain the reasons for the drift, and most scientists scoffed at his theory. After all, said his critics, earthquake waves showed clearly that, except for a molten core, the earth was solid rock. How could the continents possibly "drift" through the solid globe? And there the theory of Continental Drift rested, unheralded and unproven, buried for decades in out-of-print geology books.

Gradually, however, impressive evidence began piling up in

Above: a view down the center of the drilling derrick of *Glomar Challenger*. The ship, designed and built by Global Marine Incorporated, was ready to start work in August, 1968. The Scripps Institution of Oceanography at once arranged a voyage of discovery in the Atlantic and Pacific oceans. During two years' drilling, *Glomar Challenger* brought up cores of the ocean bed in many parts of the world. By studying the sediments contained in these cores, some of them 140 million years old, scientists have been able to answer many fundamental questions about the nature of the floor of the oceans.

support of Wegener's idea. When geologists compared the ages and positions of similar kinds of rocks in Africa and South America, they found that the layers of older rocks on both continents matched up perfectly. Identical fossils of plants and animals were found on more than one of the southern continents. The chances of perfectly similar life forms developing at exactly the same time in different places was remote. And the broad oceans that now separate the continents would be an effective barrier to most organisms. Even stronger evidence came from the study of magnetization in rocks of the same age from different continents. Scientists began to take a new look at Wegener's theory.

Meanwhile, oceanographers had been carefully plotting the course of the great mid-ocean ridge. It did not take them long to notice that, if some of the continents were joined together, the ridge would fit exactly between them. What, they wondered, did this seam around the earth have to do with the puzzle of Continental Drift?

Then, in the early 1960's, British scientists discovered the most remarkable feature of the undersea mountain range—a steep-sided valley that cut lengthwise right through the center of it. This discovery led to a new explanation of how the continents might have split apart. More important still, it suggested that the process might be continuing today.

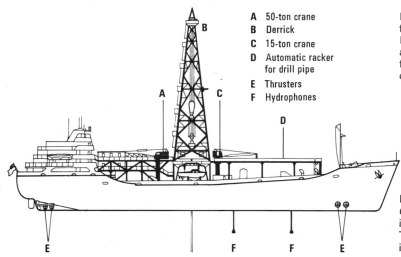

A 50-ton crane
B Derrick
C 15-ton crane
D Automatic racker for drill pipe
E Thrusters
F Hydrophones

Left: a diagram showing the main features of the *Glomar Challenger*. During drilling operations the automatic racker ensures that the 90-foot lengths of drill pipe are transferred quickly to and from the derrick.

Far left: a scale drawing of the *Glomar*'s drill pipe, when the ship is drilling in 18,000 feet of water. The actual pipe has been increased in thickness so that it shows up.

Scientists proposed that the sea floor itself was spreading apart on both sides of the crack down the middle of the mid-ocean ridge. As the crack widens and fills up with molten rock from deep within the earth, the giant plates of land that form the earth's crust are pushed aside and the continents are borne along with them.

The theory was highly attractive to scientists. But their revolutionary suggestion could only be proved or disproved by drilling through thousands of feet of sediment and into the sea floor on either side of the mid-ocean ridge. The geological evidence obtained in this way would enable them to determine whether or not the continents were drifting, how rapid was their movement, and how old the seemingly timeless oceans really are. The history of the planet seemed there for the taking by adventurous scientists. Thus, the idea of *Glomar Challenger* was born.

On November 14, 1967, Scripps Institution of Oceanography contracted with Global Marine Incorporated to design and build a radical new drilling ship that could penetrate the dark ocean floor 20,000 feet below the surface of the sea. Scripps acted as part of a consortium of five other oceanographic research institutes. Less

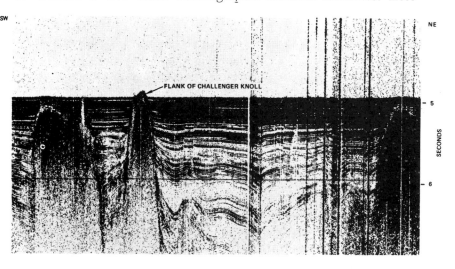

Right: a trace of the sediments below the Sigsbee Knolls in the Gulf of Mexico shows they are the tops of huge, dome-shaped formations extending thousands of feet below the sea floor.

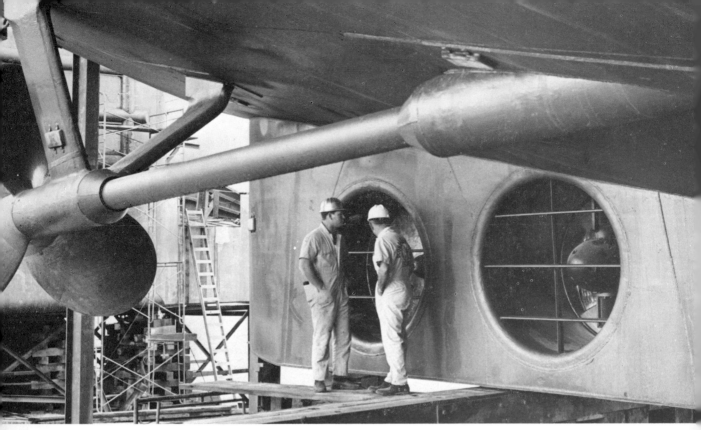

than 10 months later, on August 11, 1968, the ship was officially accepted by the group, and its epoch-making voyage began.

The *Glomar Challenger*'s first adventure came little more than a week after it had put out to sea. Near midnight on August 19, the captain rang "finished with engines" and the ship glided to a halt on the calm waters of the Gulf of Mexico. Below the keel were 11,753 feet of water—one of the deepest areas of the gulf. But it was not the depth that interested the scientists aboard the *Glomar*. On the sea floor itself was one of the greatest puzzles of marine geology, the Sigsbee Knolls. Solving the riddle of these strange, rolling hills that, in the words of one scientist, "shouldn't be there" was to be the ship's first challenge.

The Sigsbee Knolls had been discovered several years before by oceanographers from the Lamont-Doherty Geological Observatory aboard the *Vema*. The knolls lie in a broad belt 200 miles wide that crosses the Sigsbee Deep and trails away toward the southwest. Using instruments that can "see" through the blanket of sediment on the bottom, Lamont scientists had found that the knolls were only the tops of giant, dome-shaped structures that extend thousands of feet beneath the sea floor. Other instruments indicated that these vast domes might be made of salt—exactly like those found in the coastal areas of the Gulf of Mexico. These coastal salt domes often signal the presence of gas and oil and prospectors search for them eagerly.

Most geologists doubted that the domes could be made of salt. It was an accepted fact that the huge deposits of salt required to

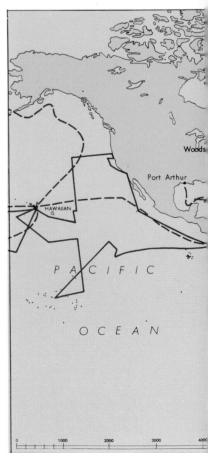

Left: the stern thrusters of *Glomar Challenger*. The *Glomar* has thrusters at both bow and stern—both consist of propellers mounted in tunnels running through the hull. They enable the ship to move sideways so that it can remain precisely over the drill site in the open sea. When drilling starts they are operated automatically as part of the dynamic position-fixing system.

Below: the route of the *Glomar Challenger* on its drilling expeditions between 1968 and 1970. By drilling into the ocean floor much can be learned about the history of the earth. The scientists of the *Glomar Challenger* have studied cores taken from the seabed and made important discoveries about the age of the oceans and about Continental Drift.

form such domes could only be laid down in shallow seas. And the Sigsbee Knolls were more than two miles down.

But at least two scientists aboard the *Glomar Challenger* were certain that the Sigsbee Knolls were indeed salt. These men were Maurice Ewing and J. Lamar Worzel, joint-chief scientists on the pioneering drilling venture of the *Glomar Challenger*. Both were experts in marine geology. Both had checked and rechecked their figures and were certain that they were correct. If the *Glomar* could successfully drill into the bottom, the solution to the riddle might be found.

When the *Glomar Challenger* halted over one of the knolls, electronic technicians went to work. Two sonar beacons were switched on. The beacons were hung over the side and then dropped free, their heavy battery cases dragging them toward the bottom at six feet per second. The sonar beacons are the heart of a dynamic positioning system that permits the *Glomar* to drill in miles of water without anchoring. Signals from the beacons are constantly picked up by four hydrophones hanging in the water beneath the ship's hull. A computer system analyzes these signals, and, as the ship drifts, automatically activates side thrusters to maintain exact position over a hole thousands of feet below.

During the 35 minutes the sonar beacons took to free fall to the

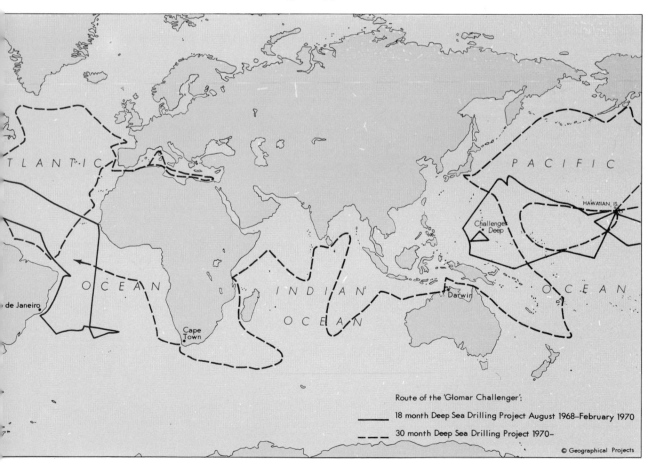

Route of the 'Glomar Challenger':

_____ 18 month Deep Sea Drilling Project August 1968–February 1970

------ 30 month Deep Sea Drilling Project 1970–

© Geographical Projects

bottom, the expert *Glomar* drillers waited patiently atop the mid-ship derrick that towers 194 feet above the waterline. Finally the go-ahead came. The beacons were down and functioning perfectly. Working through the night, the drillers attached and lowered 90-foot sections of 5-inch drill pipe through the *moon pool,* the center well on the drilling platform that opens to the sea below. No one was certain that the equipment could stand the enormous weight of the steadily lowering drill pipe. But section after section gradually probed downward toward the bottom.

Just after dawn, the drill bit reached a point a few feet above the bottom. Then the chief driller started the 300-horsepower motor assembly that turns the drill pipe from the top of the miles-long string of pipes. Seconds later, the weight indicator signaled that the drill bit had entered the sediment and was starting into the mysterious knoll.

When the bit was 64 feet into the knoll, the driller called a halt. A special coring device was lowered within the string of pipes that now dangled 11,817 feet below the floor of the drilling platform. The scientists aboard anxiously awaited the first core, and when it at last emerged, it was quickly wrapped in plastic and rushed to the laboratory. The driller started up the rig once again. More and more pipe was added to the lengthening strand.

Further cores were taken and retrieved for the scientists. The *Glomar*'s equipment is so sophisticated that they could be taken from deep inside the knoll as well as from the surface. The cores revealed that the knoll was covered with a fine sediment. Then, at 450 feet into the knoll, came a startling discovery. Traces of oil and gas were found in a core. Eleven feet further on, the bit chewed into a typical salt dome cap. It, too, contained traces of oil and gas. Ewing and Worzel were right. Salt domes, and a vast new potential source of oil, had been found on the floor of the deep sea. And, almost as important, the *Glomar Challenger* had proved that it was possible to drill deep into the sea floor while freely floating miles above on the surface of the ocean.

Since that first drill hole, the *Glomar* has drilled hundreds more. After two years of deep-sea drilling, scientists were able to state that the oceans are relatively young—the oldest not much more than 200 million years, compared with the earth's 4,500 million years. They also found very strong evidence for Continental Drift. The earth's crust appears to be spreading aside relatively rapidly. Scientists have estimated that the continents are drifting apart at a rate as high as six inches a year.

Today, the *Glomar Challenger* continues to seek out new information about the history of our planet. Her drilling system has already been improved to enable her to probe ever more deeply into the sea floor. So far, scientists from Great Britain, France, Switzerland, Italy, Australia, the Soviet Union, and Brazil have participated in her mission. Future projects will take the *Glomar Challenger* to every ocean of the world.

Above: the research catamaran *Duplus,* operated by The Netherlands Offshore Company. The range of marine activities is increasing so rapidly that new research ships are designed to be very versatile. The twin-hull design of the catamaran makes it a particularly stable platform that can be used for drilling, surveying, or diving support.

Right: the *Bannock,* the oceanographic research vessel which is operated by the National Research Council of Italy.

Right: the *Atlantis II,* the research ship operated by the Woods Hole Oceanographic Institution, U.S.A. This 280-foot-long vessel was built in 1962 to replace the ketch *Atlantis.*

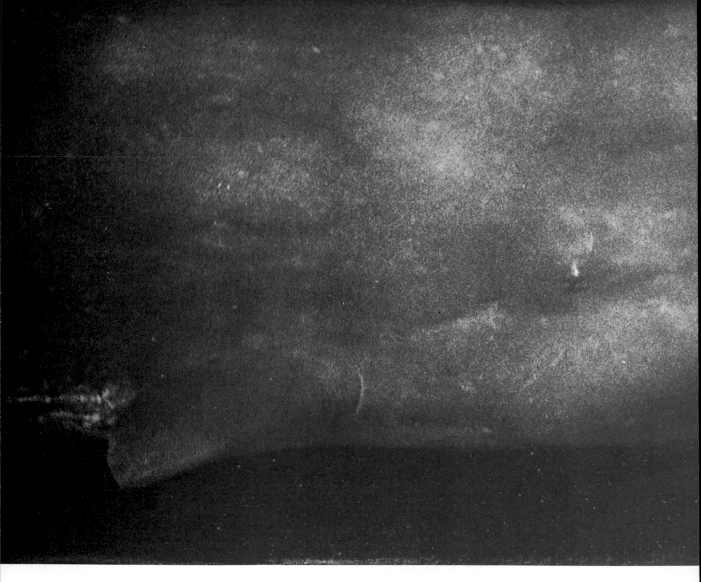

Life - Large and Small

9

There are now about $3\frac{1}{2}$ billion people in the world. At least $1\frac{1}{2}$ billion of them struggle to stay alive on a meager diet that falls desperately short of human needs. At the present rate of increase, by the year 2000, the world population will probably have almost doubled. There will be about $6\frac{1}{2}$ billion people on earth. How are these people to be fed? The sea may hold the answer, and scientists are seeking to use its resources to the full. But while the oceans abound in protein-rich fish, there is still a danger of overfishing certain species to the point of extinction. Only by careful study of marine animals and of the factors governing their life cycles can the sea's food potential be exploited to give everybody enough to eat.

During the 1800's, fishing became such a large and profitable industry that by the close of the century, stocks of some fish were

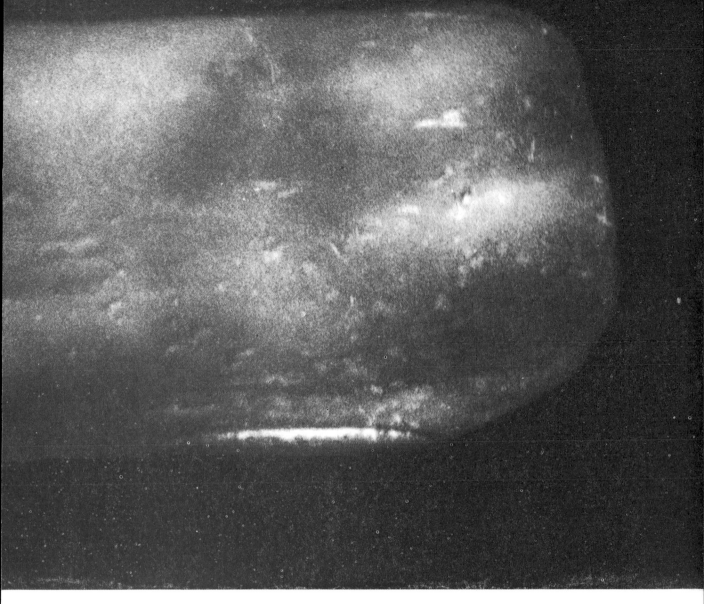

Above: a sperm whale photographed at a depth of 45 feet just off the coast of the Azores. Its broad head contains a reservoir of wax called spermaceti. During the past 100 years whales have been mercilessly hunted for their valuable supply of oil. Today, however, international restrictions on whaling are having a beneficial effect on the numbers of these superb animals.

Right: the shrimp-like crustacean that forms the food of certain baleen whales. These crustaceans, the largest members of the animal plankton, are usually called *krill.*

already seriously depleted. Governments began to realize the need for greater knowledge of marine life. Several nations organized programs of scientific research, and in 1901, the International Council for the Exploration of the Sea was set up. Its aim was to encourage governments to equip research ships and nominate scientists for the study of life and conditions in different areas of the sea.

The Scandinavian countries, whose fisheries were an important source of food, sent out a number of ships. In 1904, a Danish

Below: transparent elvers of the common eel. Eel larvae, returning from the breeding grounds in the Sargasso Sea, change into elvers three inches long when they reach river estuaries.

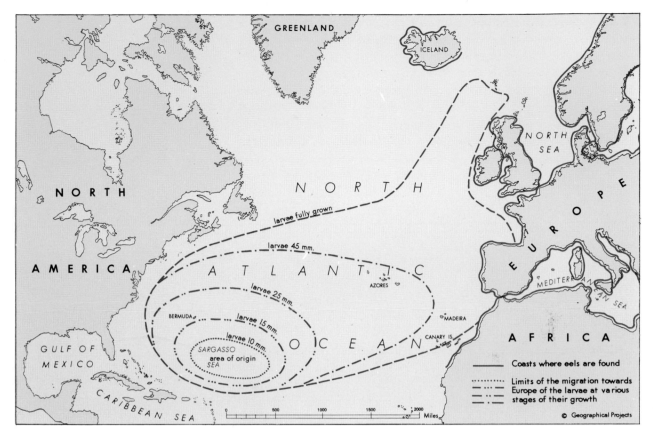

On the map:

GREENLAND

ICELAND

NORTH SEA

NORTH

NORTH

EUROPE

ATLANTIC

AMERICA

larvae fully grown

larvae 45 mm.

AZORES

larvae 25 mm.

BERMUDA

larvae 15 mm.

MEDITERRANEAN SEA

MADEIRA

larvae 10 mm.

SARGASSO
area of origin
SEA

OCEAN

CANARY IS.

AFRICA

GULF OF
MEXICO

CARIBBEAN SEA

0   500   1000   1500   2000
Miles

Coasts where eels are found
Limits of the migration towards
Europe of the larvae at various
stages of their growth

© Geographical Projects

Above: eel migration in the North Atlantic Ocean. Every year, eels from Europe travel to the Sargasso Sea, where they spawn and then die. Their eggs hatch into larvae which swim back across the North Atlantic, growing as they swim, in a journey which lasts three years. Later they, in their turn, set off for the Sargasso Sea to spawn, and so the entire cycle is repeated.

oceanographer, Johannes Schmidt, was working in the North Atlantic on the study of food fishes when he came across a creature that had puzzled and fascinated people for centuries. Schmidt's find was a *Leptocephalus*, or eel larva.

The eel has one of the most curious life stories of all fish. As long ago as the 300's B.C., Aristotle observed and dissected the eels that he found in the Mediterranean and Aegean seas, but he was unable to discover anything about their breeding habits. Aristotle noted that, at certain times, the eels migrated in great numbers to the open sea and apparently disappeared. But where did they go? And was their disappearance in any way linked with the birth of their young?

Over 2,000 years later, Schmidt began to hit on the answers. His discovery of the flat and transparent little *Leptocephalus* led him to believe that the eels' spawning ground might lie somewhere nearby. Although he was mistaken in this belief, his efforts to prove it correct enabled him to put together the true story. After 16 years of patient search in the Atlantic, he was able to piece together the final clues to the mystery of eel migrations. Aboard the research ship *Dana* in 1922, he discovered that the eels migrate to the Sargasso Sea, a large region in the North Atlantic near Bermuda. Millions of eels from America and Europe find their way to this spawning ground—and to death. Every fall, European eels leave the rivers where they have spent five or six years of their life, and swim out to sea. Making their unfailing way across the Atlantic to the Sargasso Sea, the eels show an amazing sense of direction. On arrival, they spawn and die. Their eggs hatch into larval eels (larvae) that return to Europe, growing as they travel on an amazing three-year journey back across the Atlantic. Arriving in springtime, they change into elvers (or young eels) and the female elvers swim up the rivers while the males remain in tidal waters. Several years later, the fully-grown eels set off for the Sargasso Sea again and the cycle is repeated. The same instinct drives the eels of eastern America to the Sargasso Sea to spawn, but the European eel has a longer journey.

While Schmidt was making his discoveries, other oceanographic vessels were also hard at work in the Atlantic. Among these was the Norwegian ship *Michael Sars* which made several cruises under the leadership of oceanographer and fishery scientist Johan Hjort. Hjort had developed a new type of large tow net that could be closed before being hauled to the surface. This meant that the net did not catch animals from many different layers on its way up, and gave an accurate picture of life at a chosen level in the sea.

On one of his voyages, Hjort was joined by a young British zoologist, Alister Hardy. The two men spent three weeks studying whales off Norway and Iceland. Hardy was eager to learn all he could of Hjort's methods and equipment. He was already engaged in the planning of an important mission for the British government. Soon he would set off for the icy waters of the Antarctic in a bid to save the whale from complete extermination at the hand of man.

Man has hunted the whale for centuries, chiefly for the valuable oil it yields. Whale oil used to be a major fuel for lamps and cooking. Now it is used primarily in the production of margarine and, to some extent, in soap making. The sperm whale also provides ingredients for industrial lubricants, beauty creams, and perfume. A by-product of whale oil is glycerin.

Below: *Stranded Whale,* a print by the Japanese artist Kuniyoshi (early 1800's). It shows the mixture of awe and fear with which people regarded these huge animals.

Above: tiles showing Dutch ships whaling in northern waters during the 1600's. A whaling expedition, returning from Spitzbergen in 1611, brought back news of enormous numbers of whales in the area. Competition between English, Dutch, Danes, and Biscayans then became so fierce that, in 1618, the coastline was divided up between the countries.

It was the need for nitroglycerin during World War I that greatly increased the demand for whale oil. To satisfy that demand, whalers slaughtered thousands of whales wherever they could find them. In a single whaling season, nearly 12,000 whales were killed off South Georgia alone. After the war, potential profits from whaling spurred ever more and greater attacks on the whale population of arctic and subarctic waters. Whales were threatened with extinction and so was the whaling industry.

In 1920, the British Parliament agreed on the need for a program of biological exploration to preserve the country's rich whaling industry. A team of researchers was chosen, led by Stanley Kemp and with Alister Hardy as chief zoologist.

An old and sea-scarred vessel was chosen as research ship for the scientific whaling mission. It was the *Discovery*, which had been built in 1901 for Robert Scott's first mission to Antarctica. Later, it was sold to the Hudson's Bay Company and, for many years, plied the frigid arctic waters as a cargo vessel. In 1923, the *Discovery*

Above: the British research ship *Discovery*. It was built in 1901 for Scott's first mission to Antarctica, but later, in 1923, was rebuilt for scientific research work.

returned to Britain and was almost completely rebuilt. Manned by scientists and sailors, on September 24, 1925, it sailed from southwest England toward the region of its former voyage with Scott, Antarctica. In the South Atlantic lay extensive whaling grounds exploited by Britain from bases on South Georgia and other islands of the Falkland Islands Dependencies whose facilities the *Discovery* scientists could use.

A month out at sea, the *Discovery* met with the largest school of dolphins that the crew had ever seen. Hardy was delighted by these creatures, traditionally regarded as a lucky omen by sailors. "There were certainly no less than a hundred and some estimated . . . two hundred," he noted in his journal. "Many of them kept leaping completely out of the water. . . . Who can doubt that they are leaping in play for the sheer joy of it?"

The old *Discovery* could not be hurried. She made her way slowly southward, often pausing so that the scientists could sample sea

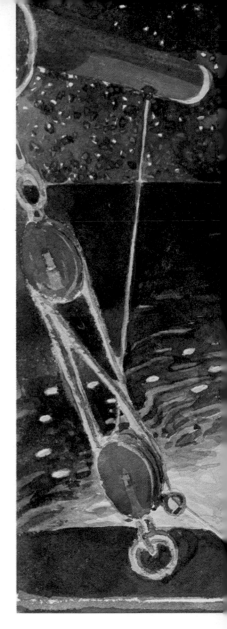

Left: a group photograph taken on board *Discovery* off South Georgia, Christmas, 1926. The director of the research program, Stanley Kemp, is fourth from the left along the seated row. Alister Hardy, chief zoologist, is sixth from the left in the same row. The photograph also includes personnel from the *William Scoresby* (a research ship) and from the Marine Biological Station at South Georgia.

water and net the living things it harbored. Fascinated by the creatures he found, Alister Hardy wrote: "Who . . . will ever forget his first catch from 2,000 or 3,000 metres' depth: the tow net bucket filled with fantastically shaped fish, often studded with luminous organs; hosts of scarlet crustaceans, deep-sea medusae, patterned and coloured like Turkey carpets, and many other creatures less easily described." Later, he reported seeing jellyfish "six to nine inches long—made of stiff jelly in which are embedded hundreds of small individual animals, each, when agitated, glowing with a bright blue-green light. For several nights after crossing the equator the ship passed through dense zones of these living lanterns, millions and millions of them, so that a broad patch of light was left behind the ship for half a mile or so."

Gradually, the colorful tropical waters were left behind. Ahead lay whale waters. *Discovery* plowed through a sea dotted with jagged

Above: a painting by Alister Hardy showing "a remarkable phosphorescent display of *Pyrosoma* in the wake of the *Discovery* on November 11, 1925, not far from Ascension Island." The crew of *Discovery* saw many such displays during their voyages in the South Atlantic Ocean. Each *Pyrosoma* (Greek *pyro,* fire; *soma,* body) is a colony of sea squirts which glow brightly, especially when disturbed.

icebergs and whipped by hurricane-force winds. Finally, on February 20, 1926, it reached South Georgia. The shore party that greeted the ship included scientists who were already in their second season of whale research. Working under the trying conditions of the whaling station, "amongst the blood, stench, and slime," they had made detailed measurements and observations of more than 1,600 whales. These included the finback whale, the humpback whale, the sei whale, the sperm whale, and the largest animal ever to inhabit the earth, the giant blue whale. (Blue whales have been claimed to measure more than 100 feet in length and to weigh 150 tons. By comparison, the largest dinosaurs are estimated to have weighed about 85 tons, and the largest elephants—the biggest land mammals now in existence—only weigh about 6 tons.)

The scientists had discovered that, with the exception of the sperm whale, these huge aquatic mammals feed on plankton and tiny shrimp-like crustaceans called *krill*. Krill- and plankton-eating whales are baleen whales. They have no teeth in their jaws, but thin plates of whalebone (baleen) extend downward from the upper jaw. The baleen forms a sieve. As the whale glides through the sea, it takes in gulps of water which it squeezes out through the baleen. Krill and plankton are trapped in bristles on the inner edge of the baleen, licked off, and swallowed. Sperm whales, however, have teeth in their jaws and feed almost exclusively on giant squids.

Alister Hardy set foot on South Georgia with high hopes of solving some of the riddles of whale behavior. He knew that whales migrate but was anxious to discover how far they travel and where they go. To track the whales' movements across miles of open sea, Hardy devised an ingenious marking method. The mark, made of silver-

Above: a school of female sperm whales. Sperm whales are found in all the oceans of the world. They can travel at 12 knots and dive to depths of up to 3,000 feet, staying down for 75 minutes at a time.

Right: the mammals of the sea, drawn to scale to show their comparative sizes. The tiny man at the left of the diagram gives some indication of the vast size of the blue, finback, and sperm whales.

plated rustless steel, was very much like a huge thumbtack. A number and instructions for its return were engraved on its flat head. The mark was mounted on a wooden arrow and shot into the whale's blubber (the thick fat under the skin) from a light shoulder gun.

The marked whales turned up in widely separated places, yielding new facts to Hardy and his colleagues. Whales, in the same way as many birds, were found to maintain migratory homes in specific areas of the sea. Blues, finbacks, and humpbacks always returned to the same breeding and feeding grounds year after year. It was discovered that whales generally mate in the winter. They then head north to warm waters, where the female gives birth about 12 months later. In the case of the blue whale, the newborn calf weighs $2\frac{1}{2}$ to 3 tons and is about 23 feet long. Feeding on its mother's fat-rich milk, the young blue puts on 200 pounds a day for about 7 months, tipping the scales after weaning at an amazing 23 tons.

Hardy and his colleagues found that whales, like many other large mammals, rarely have more than one offspring at a time. And they give birth only at two-yearly intervals. Often, this relatively slow rate of population growth is not enough to offset the numbers destroyed by whalers.

blue whale

porpoise

killer whale

fin whale

bottle-nosed dolphin

man

beluga or white whale

narwhal

sperm whale

pilot whale

ARCTIC CIRCLE

GREENLAND

ICELAND

Route of R. R. S. Discovery 1925-7
Route of 'Meteor' 1925-7

NORTH
AMERICA

EUROPE

Falmouth

N O R T H

ATLANTIC

MEDITERRANEAN S

BERMUDA

GULF OF
MEXICO
TROPIC OF CANCER

SARGASSO
SEA

OCEAN

CANARY IS.

TROPIC OF CAN

BAHAMA
IS.

VIRGIN
IS.

CARIBBEAN SEA

XIII

AFRICA

CAPE VERDE IS.

XIV

XII

IX

X

EQUATOR

SOUTH

AMERICA

SOUTH

EQUAT

XI

VIII

VI

PACIFIC

ATLANTIC

20°

TROPIC OF CAPRICORN

VII

TROPIC OF CAPRICC

II

OCEAN

OCEAN

IV

a

Cap
Tow

b

TRISTAN DA CUNHA

a

0°

I

b

a

b

III

b

a

STR. OF MAGELLAN

FALKLAND
IS.

Stanley

a

a

SOUTH
GEORGIA

Grytviken

b

V

quatorial Scale

b

500   1000   1500   2000
Miles

b

S C O T I A

SEA

© Geographical Projects

60°

Left: the routes of the *Discovery* and *Meteor* expeditions. The *Discovery* expedition studied whales and their behavior, and the tiny plankton on which whales feed. The German *Meteor* expedition (1925–1927) made observations of the properties of seawater, and other oceanographic surveys, and contributed more to the knowledge of the ocean than any previous expedition.

Right: crew members of the *Discovery* maneuver a continuous plankton recorder over the side of the ship.

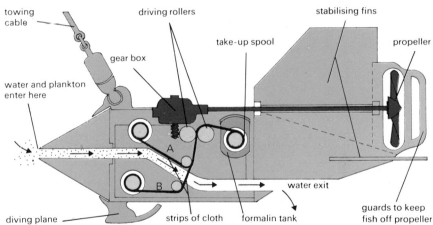

towing cable
gear box
water and plankton enter here
driving rollers
take-up spool
stabilising fins
propeller
A
B
diving plane
strips of cloth
formalin tank
water exit
guards to keep fish off propeller

Right: a diagram of the redesigned plankton recorder used on later voyages. As the device moves through the water, plankton is trapped on a roll of silk netting moving across the water channel. The entangled specimens are covered and rolled into a preserving tank full of formalin.

After weaning, mother and calf begin their journey to feeding grounds a thousand or more miles to the south. On the way, the calf gets its first taste of krill and plankton—the foods that will sustain it for the rest of its life. Hardy knew that if the survival of whales was to be guaranteed by man, the biology of the tiny krill would have to yield its secrets to the *Discovery* scientists.

Until the voyage of the *Discovery*, no one had paid much attention to the life history of krill. Hardy had to start from scratch. First, he invented a device, which he called a "continuous plankton recorder," that could snare krill and plankton over long distances and at various depths. The Hardy plankton recorder, as it has come to be called, probed depths up to 600 feet. It revealed that the sea is not covered with a continuous and uniform blanket of plankton, but has patches here and there that the whales must search out in order to survive. As the larger whales can consume as much as $1\frac{1}{2}$ tons of food a day the supply of plankton is extremely important. The impact of this observation was vividly demonstrated to Hardy one summer day as the *Discovery* plowed through the cool waters of the Falkland Islands Dependencies. Just beneath the surface of the water a five-mile-patch of red krill-containing plankton stood out like a beacon beckoning the whales. And they came—150 to 200 whales in search of food. Oblivious of the ship, they strained the plankton from the sea and later played in the sea like puppies, sending spouts of moisture-laden air shooting from their blowholes.

Hardy's findings did not bring about immediate action to curb the slaughter of whales. About 1 million whales have been killed in Antarctic waters since he made his historic voyage. Only recently have whaling nations agreed on adequate protective measures. But to Alister Hardy and whaling scientists like him must go much of the credit for making these reforms possible.

At the same time as the *Discovery* scientists were making their investigations, an important German expedition was also at work in the South Atlantic Ocean. The German team, aboard the *Meteor,* crisscrossed the South Atlantic, making detailed studies of the temperature and chemical composition of sea water. These observations, together with geological and meteorological research carried out from the *Meteor,* contributed a great deal to basic understanding of the ocean.

While knowledge of the ocean itself was increasing, scientists continued to be intrigued by the creatures that live in the underwater world. Hardy had studied the giants of the sea and the plankton upon which they feed. But far below the surface regions populated by plankton, roamed other creatures, as yet undiscovered. It was in search of these that the Danish oceanographic research vessel *Galathea* steamed from Copenhagen on October 15, 1950.

When the *Galathea* expedition was planned, scientists were unsure whether animals lived at depths greater than 19,500 feet. According to Anton Bruun, the scientific leader of the expedition, "The primary purpose of the *Galathea* expedition was to explore the ocean trenches in order to find out whether life occurred under the extreme conditions prevailing there—and if so, to what extent."

For Bruun the voyage was doubly important. As a child, his dearest ambition had been to become a sailor. But this dream could never be realized. Before he was 10 years old, poliomyelitis had left him permanently lame. He could not forget the sea, however, and began a study of marine animals that eventually won him a place as assistant to Johannes Schmidt aboard the *Dana*. Then, at last, the *Galathea* gave him his chance to turn the tables on destiny. The man who could not be a sailor would wrest the secrets of the sea from the deepest ocean abysses.

On the night of July 22, 1951, the *Galathea* was in the Pacific Ocean about 125 miles northeast of the island of Mindanao in the Philippines. Her echo sounder had located a hole 33,678 feet down in the Mindanao Trench, a 540-nautical-mile crevice in the ocean floor. For 110 minutes *Galathea*'s sledge-trawl was dragged along the ocean bottom over six miles down. It was the largest trawl

Above: the water turns red as the men of the Faeroe Islands close in on a school of pilot whales. The islanders drive the 10- to 25-foot-long whales toward shallow water before attacking them with harpoons and knives. Today this massacre is a festival: in former times the whale meat it provided often meant survival through the winter.

Left: a sei whale (foreground) and a finback whale on the deck of a whaling boat. Both are baleen whales whose principal food is plankton.

to reach such a depth and no one knew what it would bring up.

It took several hours to haul the trawl to the surface. Oceanographers and crew gathered around eagerly as the net was opened. Sea anemones, sea cucumbers, clam-like shellfish, an *amphipod* (a burrowing crustacean), and a bristle-worm spilled out among the stones and mud from the deep ocean floor. A whitish sea anemone was discovered which had never before been seen by man. Bruun had found a variety of life in the pressure, darkness, and extreme cold of one of the greatest depths of the ocean.

During the *Galathea*'s two-year voyage in the Pacific, the Caribbean, and the Atlantic, many extraordinary new creatures were discovered. Deep-sea animals depend for food on the life in the ocean layers above them. The open sea is much less rich in life than the area around the coasts. This means that abyssal creatures living a long

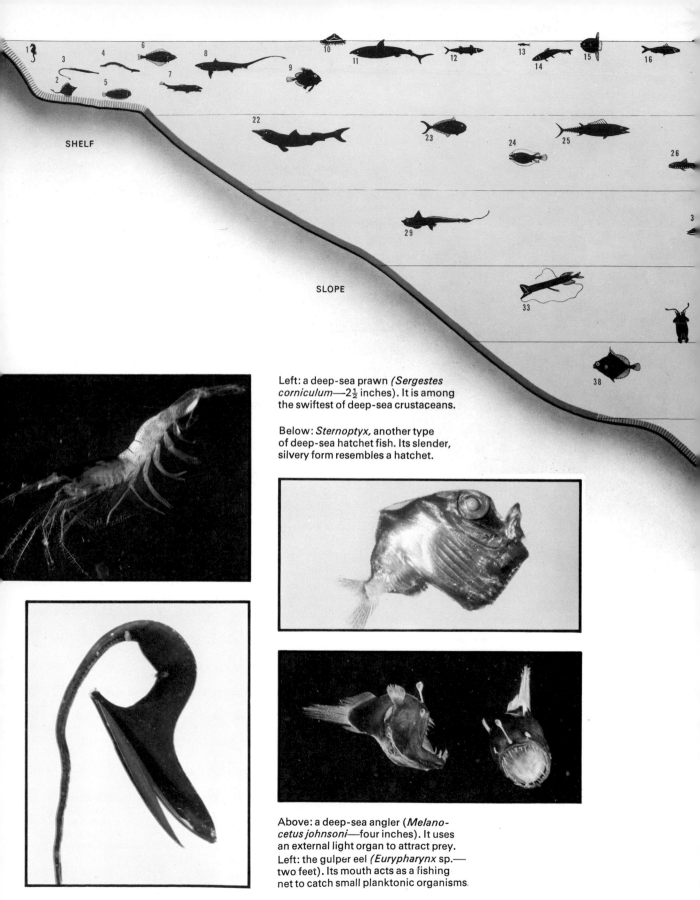

SHELF

SLOPE

Left: a deep-sea prawn *(Sergestes corniculum—*2½ inches). It is among the swiftest of deep-sea crustaceans.

Below: *Sternoptyx,* another type of deep-sea hatchet fish. Its slender, silvery form resembles a hatchet.

Above: a deep-sea angler (*Melanocetus johnsoni—*four inches). It uses an external light organ to attract prey.
Left: the gulper eel *(Eurypharynx* sp.— two feet). Its mouth acts as a fishing net to catch small planktonic organisms.

0 feet	
600 feet	
1200 feet	
1800 feet	
2400 feet	
3000 feet	
3600 feet	
4200 feet	
4800 feet	
5400 feet	
6000 feet	

ABYSS

Left: the diagram shows some of the swimming animals of the Atlantic. In the deep ocean, where no light penetrates, many animals have light-producing organs, and some are grotesquely adapted to their environment.

1. Sea horse (*Hippocampus europaeus*) 7 inches
2. Common skate (*Raja batis*) 6 feet wide
3. Broad-nosed pipefish (*Syngnathus typhle*) 12 inches
4. Common eel (*Anguilla anguilla*) 2 feet 6 inches
5. Sole (*Solea solea*) 12 inches
6. Halibut (*Hippoglossus hippoglossus*) 3 feet
7. Cod (*Gadus callarias*) 2 feet 6 inches
8. Thresher shark (*Alopias vulpes*) 15 feet
9. John Dory (*Zeus faber*) 8 inches
10. By-the-wind-sailor (*Velella velella*) 2 inches
11. Basking shark (*Cetorhinus maximus*) 25 feet
12. Mackerel (*Scomber scombrus*) 12 inches
13. Pilchard (*Sardina pilchardus*) 8 inches
14. Herring (*Clupea harengus*) 9 inches
15. Sunfish (*Mola mola*) 8 feet
16. Allis shad (*Alosa alosa*) 9 inches
17. Portuguese man-of-war (*Physalia physalis*) 10 inches
18. Flying fish (*Exocoetus volitans*) 9 inches
19. Sperm whale (*Physeter catodon*) 50 feet
20. Common dolphin (*Delphinus delphis*) 7 feet
21. Black fish (*Centrolophus niger*) 3 feet
22. Blue shark (*Prionace glauca*) 15 feet
23. Ray's bream (*Brama raii*) 1 foot 6 inches
24. Angler (*Haplophryne* species) 1 ½ inches
25. Bluefin tunny (*Thunnus thynnus*) 8 feet
26. Lantern fish (*Myctophum punctatum*) 4 inches
27. Hatchet fish (*Argyropelecus affinis*) 4 inches
28. Sea robin (*Peristedion miniatum*) 2 feet
29. Ratfish (*Chimaera monstrosa*) 3 feet
30. A stomiatoid fish (*Cyclothone pallida*) 2 inches
31. Prawn (*Acanthephyra multispina*) 3 inches
32. Devilfish (*Linophryne arborifera*) 3 inches
33. A stomiatoid fish (*Ultimostomias mirabilis*) 2 inches
34. Little post-horn squid (*Spirula spirula*) 2 inches
35. A stomiatoid fish (*Bathophilus melas*) 12 inches
36. Giant squid (*Architeuthis princeps*) 40 feet
37. Frilled shark (*Chlamydoselachus anguineus*) 5 feet
38. *Cyttosoma helgae* 9 inches
39. Bat fish (*Malthopsis erinacea*) 5 inches
40. Squid (*Desmoteuthis pellucida*) 5 inches
41. Viperfish (*Chauliodus sloani*) 12 inches
42. Snipe eel (*Cyema atrum*) 12 inches
43. Cross-toothed perch (*Chiasmodus niger*) 5 inches
44. Giant-tail (*Gigantura chuni*) 12 inches
45. Wonder-lamp squid (*Lycoteuthis diadema*) 2 inches
46. Big-headed rat-tail (*Macrourus globiceps*) 12 inches
47. A stomiatoid fish (*Cyclothone microdon*) 3 inches
48. Prawn (*Sergestes corniculum*) 2½ inches
49. Squid (*Histioteuthis bonelliana*) 2 feet
50. Oar fish (*Regalecus glesne*) 12 feet
51. Gulper eel (*Eurypharynx pelecanoides*) 2 feet
52. Lantern fish (*Lampanyctus pusillus*) 4 inches
53. Angler fish (*Melanocetus johnsoni*) 4 inches
54. Hatchet fish (*Sternoptyx diaphana*) 2½ inches
55. Angler fish (*Gigantactis macronema*) 5 inches
56. Devilfish (*Caulophryne acinosa*) 1 ½ inches

Drawings not to scale. Approximate sizes are given.

distance from the shore often have special adaptations for obtaining food as well as for coping with the other difficulties of their environment. The black swallower, for example, can eat fish up to three times its own size. Others, like the *Stylephorous,* have mouths that shoot out rapidly in front of their heads, or mouths shaped like scoops for digging into the mud in search of food.

Most deep-sea fish belong to the *Brotulidae* family, and the *Galathea's* nets dredged up large numbers of these blind transparent little creatures. They also found members of the *Mac-*

Above: drawings showing the deep-sea viperfish *(Chauliodus sp.—* 10 inches) engulfing prey by swinging its jaws outward and upward. This process allows the fish to swallow animals larger than itself—a useful adaptation to life in the depths of the sea where food is in short supply.

Right: a deep-sea angler fish *(Chaulophryne jordani*—six inches). A parasitic copepod (small crustacean) is attached near the angler's tail.

rouidae, or rat-tail, family who, unlike the brotulids, have sharp eyes. This is because they spend their early days in the light, upper levels of the sea before moving to the perpetual darkness of the deep.

Some animals, however, carry their own portable light with them. A certain type of prawn called *Acanthephydra* produces a mass of "living light" from pores beneath its eyes. The squid also emits a cloud of light in a similar way to an octopus squirting ink. These lights may serve to confuse an enemy, or for communication between animals of the same species. But they can also be used for hunting.

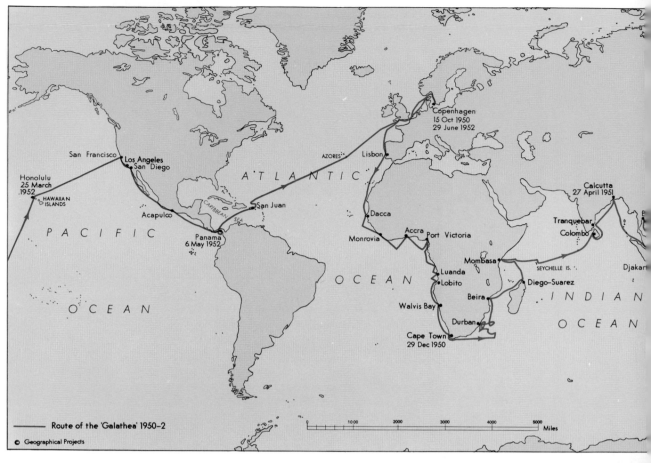

Route of the 'Galathea' 1950-2

© Geographical Projects

This use is dramatically demonstrated by the deep-sea angler.

A plump, black, scaleless creature, the deep-sea angler is relatively small but its jaws are armed with many long, inward-curving teeth. Just above its mouth is an arch-shaped light organ—the angler's special hunting apparatus. When a small fish, attracted by the light, ventures too close, it is instantly snapped up in the angler's jaws.

Up until May 6, 1952, this external light organ was the only kind known to be used by the many species of deep-sea anglers. On that day, the *Galathea*'s trawl was scraping along the ocean bottom off the west coast of Central America in waters more than 11,500 feet deep. There it trapped a giant, 18½-inch-long angler. Its size, startling as it was to the scientists, was the least of the surprises presented by the fish. It had no light organ on its head! Where was its "lure"? Had it been knocked off when the animal had been captured? A look inside its gaping mouth revealed the answer.

There, behind the menacing teeth and suspended from the roof of the mouth, was a fork-shaped light organ. The *Galathea* zoologists guessed that this creature rests on the ocean bottom, its mouth open, its light glowing, simply waiting for its prey to dart in. The *Galathea* trawl had brought up one of the laziest hunters of the deep.

Man sometimes takes a tip from the fish world in his own methods

136

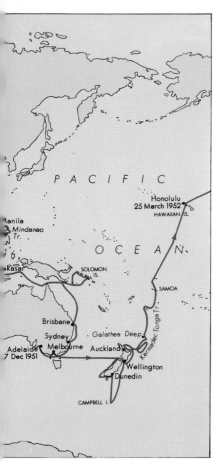

Left: the voyage of the *Galathea*. According to Anton Bruun, the leader of the *Galathea* expedition, the purpose of the expedition was "to explore the ocean trenches in order to find out whether life occurred under the extreme conditions prevailing there—and if so, to what extent." The *Galathea* answered this question when it trawled creatures from the sea-bed 33,678 feet below the surface.

of hunting food from the sea. It is not unusual for fishermen to use lights to attract fish to their waiting nets. Modern techniques combine this light lure with the use of a suction pump instead of a net. The pump draws the fish up from the sea into the ship's hold.

The fishing industry has been revolutionized in recent years by the invention of technological devices to locate rich fishing grounds, to improve the size and quality of catches, and to process the fish. With these and more traditional methods, the world's fish catch has trebled in the last 20 years to reach an annual total of around 64 million tons. It has been estimated that this could increase to as much as 200 million tons by 1985. But, at the same time, overfishing is still having a disastrous effect on some stocks of fish and will lead to a

Below: the boat in the center has caught a huge shoal of herring with a purse seine. The net was towed around the shoal and then drawn together at the bottom to form a bag.

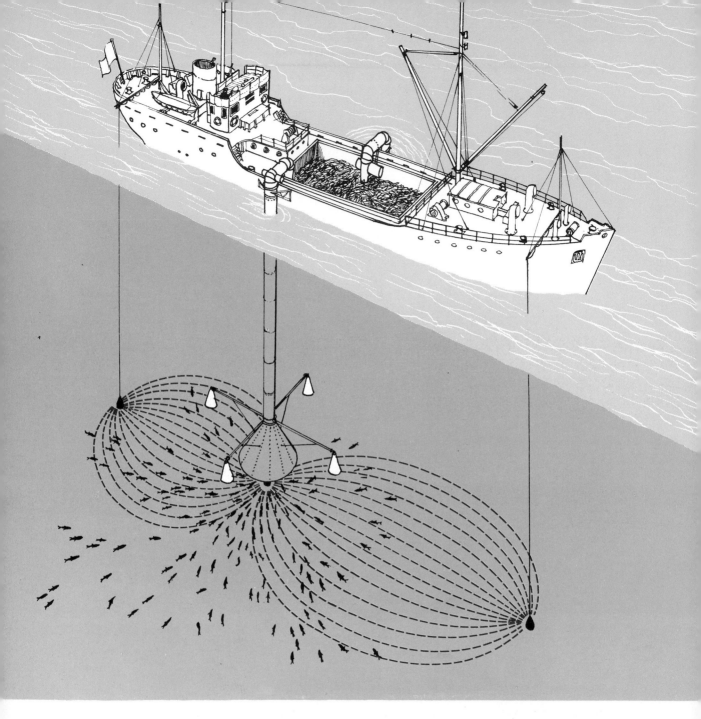

Above: fishing without nets. The fish
are attracted by strong underwater
lights toward the funnel-shaped
opening of a suction pipe. Electrodes
suspended in the water set up an
electric field, stunning the fish, which
are pumped up into the ship's hold.

Right: an underwater fish farm of the
future. By the year 2000 the sea will
have to produce one-fifth of the world's
food. The drawing shows how auto-
mated rearing tanks built on the bottom
of the sea may make this possible.

138

decline in future supplies if it is not checked. Just as in the case of the whale, protective measures must be worked out on the advice of marine biologists who are now doing a great deal of research into this problem.

In the meantime, fish farms are providing a substantial new source of food from the sea. These farms work in several different ways. Fish may simply be caught in the ordinary way and held in enclosed areas in or near the sea, for sale when other stocks run low. Alternatively, young fish may be taken from the sea and fattened in enclosures until they are ready for the market. The Japanese have been particularly successful with this technique, producing 20,000 tons of a species of tuna and 13,000 tons of eels from their farms in the course of a single year.

Yet another method involves transferring young fish from overcrowded fishing grounds to fresh areas where there is plenty of food for them. A more advanced technique is to grow fish from eggs in special hatcheries. Once the fish are hatched, they may be released into the open sea to increase existing stocks, or reared to marketable size within the fish farm, where they can be protected from predators and diseases. When the whole process takes place under the controlled conditions of a fish farm, there is also the possibility of breeding new strains of fish.

One of the most successful experiments in marine farming to date is the rearing of prawns. Here again, Japan leads the world with a record 300–400 tons of prawns farmed a year. The Japanese also have extensive oyster farms, capable of producing annually 46,000 pounds of protein-rich sea food per acre of oysters. This is about 200 times as much as the meat production from cattle per acre of pastureland.

The science of aquaculture, as fish farming is known, is being actively developed all over the world. And aquaculturists are not confining their efforts to foods already accepted as "edible" by most of the world's people. They are investigating the possibility of growing and harvesting seaweeds and various kinds of plankton—all rich in vitamins, minerals, and protein. Scientists estimate that the sea's 89 billion acres produce more than 400 billion tons of plant plankton alone during a year. Compared with the 64-million-ton annual fish catch, this represents a huge amount of plant food that might be used to feed the earth's hungry people if such food could be efficiently harvested and made palatable. Some scientists are at work

Above: fish being pumped from the sea. Russian fishermen in the Caspian Sea use this technique to catch sardines. Such methods, together with the development of new nets and fish-finding equipment, have boosted the world fish catch, but they may lead to over-fishing of certain vital food fishes.

trying to develop a plankton flour for making bread or crackers.

In the future, many of our medicines, too, may come from the sea. Scientists have discovered ingredients of certain algae that can be made into highly beneficial drugs. A few of these drugs are already on the market and others are being eagerly sought. One danger of today's widespread use of antibiotics is that bacteria may gradually build up a resistance to them and make them ineffective. This makes it vital for scientists to find new antibiotics.

The American botanist Paul Burkholder has made a detailed study of algae during recent years. Working on 6 species of seaweed, he has discovered at least 15 antibiotic substances. Other specialists have found antibiotics in shellfish such as the clam. Insulin has been extracted from starfish. And the octopus has yielded medicine for the treatment of certain heart diseases. Experiments have shown that some seaweed extracts can prevent the formation of blood clots.

Over 100 years ago, French economist Eugene Noel wrote: "The ocean can be turned into an immense food factory. It can be made into a more fruitful laboratory than the earth. Fertilize it! Seas, rivers, and ponds! Only the earth is cultivated. Where is the art of cultivating the waters? Hear, ye nations!"

Today, the nations have begun to listen—they have no option.

The Conquest of Inner Space

10

Left: this rabbit is in a cage that is completely submerged in water. The air that the rabbit breathes is passing through a membrane stretched over the sides of its cage. The membrane prevents water entering but passes oxygen and carbon dioxide as does the lining of the lung. Diving experts consider that such membranes could be used to construct artificial *gills* for breathing underwater.

Man has come a long way in his understanding of the world oceans since Aristotle first cataloged the marine creatures of the Mediterranean more than 2,000 years ago. Primitive diving gear and submarines have evolved into the Aqua-Lung and undersea submersibles capable of penetrating to the deepest parts of the sea. Life in the ocean, once thought to be confined to the shallow shelves surrounding the land, is now known to exist even in the greatest depths. The great currents, rivers in the ocean, have been charted and measured. The dynamic, constantly changing nature of the sea floor is known, and the history of our planet and its oceans is becoming clearer.

What, then, does the future hold for the exploration of the sea? In 1870 Jules Verne wrote in *Twenty Thousand Leagues Under the Sea,* "One must live within the ocean." And that prediction, now more than 100 years old, seems to point the way toward the next great era of ocean exploration. Man will become—perhaps *must* become—a dweller in the sea.

The first tentative steps toward living in the sea and tapping its riches have already been taken. In the late 1950's and during the 1960's, divers from Europe and the United States pushed the depth limit for diving to 1,000 feet and set up records for time spent underwater. From these achievements came new knowledge about the two greatest problems faced by the deep diver—nitrogen narcosis and the need for decompression.

It was found that, when certain inert, light gases were used,

Right: Waldemar Ayres demonstrates his artificial *gills* on New York's Jones Beach. Each gill consists of two plastic sheets, at the bottom of which is a membrane permeable to gases but not to water. Ayres stayed underwater for 1½ hours, inhaling oxygen directly from the seawater.

nitrogen narcosis did not occur or was greatly reduced. The chief gas now supplied to deep divers is helium. This is combined with a small percentage of oxygen to form an oxy-helium mix suitable at depths of at least 800 feet. Below this depth, however, even this mixture becomes appreciably less easy to breathe. Experts are studying the possibility of replacing helium with hydrogen—the lightest of all known gases. But hydrogen has the disadvantage of being highly flammable and dangerous to handle. Nevertheless, in view of the progress which has been made to date, experienced

Left: the *Seachore* submersible diving chamber, with a decompression chamber behind it. On a routine dive, two divers are sealed in the chamber and winched down to the work site. The chamber is then pressurized to the outside water pressure and the divers leave through a hatchway. When the job is complete, the divers seal themselves into the chamber and return quickly to the surface. There they transfer to a decompression chamber, either to decompress slowly or to wait at pressure until the next dive.

Below: the *Cachalot* diving chamber. This chamber, developed by Westinghouse Underseas Division, is used at depths of below 600 feet. Once at pressure the four-man crew works alternate two-man shifts, remaining in a pressurized chamber between shifts.

divers have predicted dives to 3,000 feet within the next 20 years.

Divers who have taken part in long stays underwater have made another significant discovery, known as *saturation diving,* that helps cut wasteful decompression time. The amount of gas that the body tissues can absorb is limited. Once the saturation limit is reached the gas in the tissues does not increase no matter how long the diver remains at the depth where saturation occurs. As decompression time depends on the degree of saturation, once saturation point is reached, the diver can prolong the time spent at depth for days, weeks, or longer, without increasing the decompression time needed afterward.

Despite the use of saturation diving techniques, decompression is still a lengthy process. A diver might need only 3 minutes to reach a depth of 200 feet, but it will take him more than $2\frac{1}{2}$ hours to return to the surface, decompressing as he comes up. Decompression time cannot be reduced without danger to the diver. This presented a major obstacle to companies wishing to use divers in commercial

operations underwater. But how could the diver be brought up rapidly without cutting decompression time?

The answer came from a device known as a submersible diving chamber (SDC) that takes the diver down and carries him back to the surface under pressure. Among the SDCs now in use is Divcon International's *Seachore,* a single chamber that can take two divers down to 600 feet. Inside the chamber, cylinders supply the necessary gas for the divers and for the chamber itself. The divers leave the chamber through a hatchway and return when their work is finished. The *Seachore* is then brought to the surface and locked onto a deck decompression chamber (DDC) to which the divers transfer. The divers may wait in the pressurized DDC until they are needed for further work below. When all tasks are completed, the divers undergo full decompression.

The SDCs provide divers with a valuable work base. But if divers are to work underwater for weeks or months at a time, they need a "home" where they can live in warmth and comfort for the duration of their stay. Since 1962, when Cousteau set up his first underwater habitat, *Conshelf I,* developments in this field have been carried out almost exclusively by American and French teams.

While Cousteau was making his experiments with *Conshelf,* George Bond of the U.S. Navy had been carrying out a series of experiments with divers using pressure simulators in the laboratory to approximate to conditions underwater. The success of this work

Above: the first team of aquanauts to live in the U.S. Navy's *Sealab II* habitat. Second from the left in the front row is ex-astronaut, Commander Scott Carpenter, U.S.N., (NASA).

Above: an artist's impression of the United States Navy habitat *Sealab III* on the seabed. This habitat is designed to operate at 610 feet and to support five 8-man teams for periods of 12 days. Cables from the shore provide power, fresh water, and emergency communications. Breathing gas and emergency supplies are provided by a surface support ship. During their stay on the ocean floor, the divers test diving suits, communications equipment, and underwater tools.

led to the setting up, in 1964, of the U.S. Navy's Project Sealab.

Sealab I, a cigar-shaped cylinder, 35 feet long and 12 feet in diameter, was sited off the coast of Bermuda. There, 4 men lived for 10 days at a depth of 192 feet, swimming out often to work in the sea. Communications and other facilities reached them through cables from the shore, while breathing gas and emergency supplies were provided by a surface support ship.

The following year, *Sealab II* went into operation at a depth of 206 feet. This was an ambitious project involving three 10-man teams who took turns at staying in the habitat for 15-day periods. Among the men who took part was astronaut Scott Carpenter. He took to inner space so well that he stayed down for 30 consecutive days. Then, in spring 1969, *Sealab III* was developed to operate at 610 feet. For the *Sealab III* mission, divers from Britain, Canada, and Australia were invited to join the five 8-man teams or to act as observers.

The most recent American underwater habitat is *Tektite II,* which consists of two 18-foot high cylinders joined by a tunnel. Within the cylinders are a laboratory, crew quarters, a control center, engine room, and *wet room* (the room by which the divers enter and leave the habitat). A large freezer supplies the crew's store of food. The habitat contains a pressurized breathing mixture and is

kept at a temperature of 80°F. A small, two-man *minitat* is also used on the project for work at greater depths. The aquanauts are linked to the surface by television and radio, and surface observers keep a close check on their health and reactions during the long periods spent under confined conditions in an isolated and hostile environment.

One of the difficulties faced by the aquanauts is that of keeping warm. The habitats are well heated but the sea outside is extremely cold. This means that the aquanuats must wear specially insulated diving suits, known as *wet suits*. These are made of a foam material and may be worn over long woolen underwear. A film of water is allowed to circulate between skin and suit. The film of water is warmed by the body and acts as an insulator. Below 200 feet, however, the pressure of the water outside squashes the suit flat against the skin and insulation is lost. The *Conshelf III* divers tried to overcome this problem by wearing incompressible vests made of glass "micro-balloons" filled with carbon dioxide. But these vests were too rigid to allow freedom of movement. Other types of suits are now being tested, some incorporating electrically-heated undersuits or involving an auxiliary heater carried by the diver.

Communication is another problem for the habitat dwellers. The large amounts of helium used in the breathing gas mixtures distort

Above: inside the U.S. Navy habitat *Sealab II*. John Reaves fills out one of the many logs the aquanauts kept during their stay in the habitat, 206 feet deep in the ocean. The light spilling into the sea from the port-holes has attracted a shoal of fish.

normal speech. The aquanauts gradually become accustomed to this but it slows communication considerably, particularly between the habitat and the people on the surface. Speech deciphering equipment is helping to solve this difficulty.

The greatest obstacle still to be overcome is that of decompression. At the end of their stay in the habitat, the aquanauts have to spend as many as 20 hours in a decompression chamber. A revolutionary solution to this problem came from pioneer aquanaut Jacques-Yves Cousteau. He proposed nothing less than the creation of a new species of man—underwater man, or *Homo aquaticus*. This man, suggested Cousteau, would breathe underwater as naturally as a fish. He would be able to descend to the ocean depths without breathing gas and return to the surface without decompressing. All the equipment *Homo aquaticus* would need would be a small life-support machine plugged directly into his circulatory system. His lungs would be filled with an incompressible fluid and this is what he would breathe.

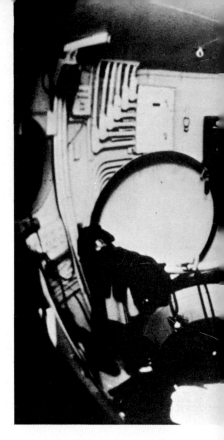

Cousteau's sensational prediction was made as long ago as the early 1960's. Already considerable advances have been made toward turning his prophesy into reality. A number of experiments with liquid-breathing animals were carried out by Johannes A. Kylstra at the University of Leiden in The Netherlands. In 1963, Kylstra went to the United States and joined with American scientists to make further tests with mice and dogs. Five years later, he was able to report on his first experiment with a human volunteer, deep-sea diver Francis J. Falejczyk. The air in one of Falejczyk's lungs was replaced by a saline solution that was "breathed" by being regularly added and drained off in equal measure. Falejczyk said afterwards that he had felt no discomfort whatsoever. Breathing with the liquid-filled lung had felt no different from breathing normally.

At about the same time, bathers on New York's Jones Beach were witnessing another remarkable experiment in underwater breathing. New Jersey diver and inventor Waldemar Ayres was trying out his synthetic *gills*. With them, he hoped to extract oxygen from the seawater and breathe under the sea just as fishes do. The gills consisted of two plastic sheets on the bottom of which was a *permeable membrane*. The membrane allowed oxygen to pass into the space between the sheets, but kept water out. Connected to this space was a system of rubber hosing through which Ayres could take in the oxygen and breathe out carbon dioxide that would be released into the seawater. Using this apparatus, Ayres stayed underwater for $1\frac{1}{2}$ hours, inhaling oxygen directly from the sea. His breathing came easily. His gills worked perfectly.

Looking into the future, Cousteau has predicted that *Homo aquaticus* will be able to move freely at depths down to about 6,000 feet. He even sees the possibility of a new race of men being born in the depths of the sea and living in huge underwater cities.

The techniques for building small undersea colonies, at least, are already within our grasp. One large corporation has completed an

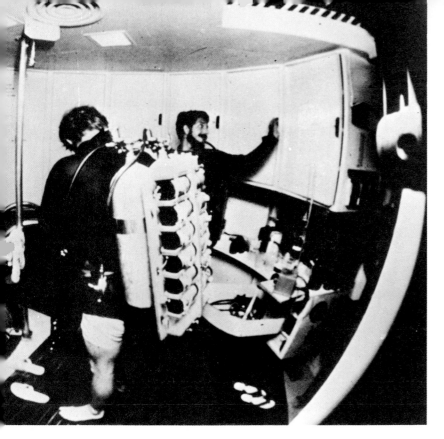

Left: a diver wearing a hot water suit supplied by an umbilical cable. Keeping warm in the sea is one of a diver's main problems. Other suits include one with a closed hot water system and one heated by electricity.

Above: divers prepare to swim out from the *wet room* of *Tektite II*. This American habitat, established during summer, 1970, has provided information on the psychological reactions to complete isolation.

Above: a diver with a powered under-water wrench. Divers underwater can-not exert pressure with normal tools. Below: a diver using a rebreather Aqua-Lung. Exhaled carbon dioxide is scrubbed so that air can be used twice.

Left: a full-size training model of the Deep Submergence Rescue Vehicle (DSRV) being loaded onto a C-141 aircraft. The DSRV is designed to rescue survivors from wrecked submarines in depths of up to 5,000 feet. It can be flown quickly to the disaster area and taken down to the stricken vessel on another submarine.
Below: an artist's impression shows how the DSRV rescue skirt fits over a submarine's escape hatch. The DSRV can rescue 24 survivors at a time.

engineering study of a sea-floor complex that could house 50 men at depths as great as 1,000 feet. The three-story, starfish-shaped structure is connected with the surface by a pressurized *elevator* hanging from a surface vessel. Cost of the habitat is estimated at about 10 million dollars. Although such habitats are designed for scientific and commercial research, it may not be long before some of them become tourist centers too.

As these undersea hamlets are developed, new kinds of underwater vehicles, called submersibles or submarinos, are also being built for manned and unmanned exploration of the deep. Some of these are designed as scout vessels to reconnoiter the millions of square miles of ocean floor that are still unknown. Others are workboats intended to operate hundreds and thousands of feet down. Future submersibles will act as sub-surface *buses* to transport workers or tourists to and from undersea habitats.

Since 1960, when Piccard and Walsh made their historic dive in the bathyscaph *Trieste,* a large number of submersibles have gone into operation all over the world. More than 40 of these are American. American interest in submersibles, however, came less from the *Trieste's* success than from a disaster which occurred in 1963.

On April 9, 1963, the United States Navy nuclear submarine *Thresher* sank in 8,400 feet of water. High above on the surface, the crew of a rescue ship stood helplessly by. They could hear the

Thresher breaking up but were unable to go to her assistance. No system existed to save submarine crews at depths below 600 feet. All the men of the *Thresher* perished.

Reaction to the disaster was swift. The U.S. Navy set about obtaining funds for one of the most ambitious oceanographic projects ever known. In 1964, they established the Deep Submergence Systems Project to build a fleet of manned submersibles which could be flown anywhere in the world within hours, and would be capable of rescuing submariners from depths of up to 5,000 feet. Though this

Right: *Beaver IV*. This submersible is
one of a series built for the offshore
oil and gas industry. Using its manipulators it is able to carry out maintenance work on undersea wellheads
at a depth of 2,000 feet underwater.
Below: the free-swimming submersible
Deepstar 4000. Since its first dive
in 1966, this submersible has been
used for many oceanographic studies.
The 18-foot-long steel craft carries
a crew of 3 to a depth of 4,000 feet,
traveling at speeds up to 3 knots.

project could not have helped the men of the *Thresher,* in saving the
lives of submariners from these relatively shallow depths techniques
may be developed which can be used for deep depth rescues.

Unmanned vehicles may also be used in rescue operations. Controlled from the surface, these vehicles are known as *telechirics,* from
the Greek words *tele* (meaning far) and *cheir* (meaning hand). One of
the most famous telechirics is the American Cable-controlled Underwater Recovery Vehicle (CURV). This device, used in conjunction
with the manned submersibles or submarinos *Alvin* and *Aluminaut,*
played an important part in the recovery of an H-bomb that was lost
in 2,850 feet of water off the coast of Spain in 1966.

One of the most adaptable modern underwater vessels is the
Deepstar 2000. The *DS-2000,* as it is known, is one of a family of
Deepstar undersea vehicles belonging to Westinghouse Electric
Corporation. It measures 19 feet long and is built of steel and glass-
reinforced plastic. Two men can travel for 45 hours in the *DS-2000*
using a self-contained life support system. After taking the submersible down, the men pilot it while lying almost flat on specially
adjusted couches.

Inside the *DS-2000* is a wealth of advanced technological equip-

Left: the rig *Orion* burning off
natural gas in the North Sea. More
than 700 drilling rigs are now probing
the seabed for oil or gas. By 1980
a third of the world's oil supply
will come from beneath the ocean.

ment. Outside are two sets of mechanical arms, capable of lifting weights up to 50 pounds. These manipulators can collect samples from the ocean bed and put them into a basket, or set up instruments for the crew.

The *DS-2000,* like nearly all existing submersibles, is powered by lead-acid batteries. These are extremely heavy and occupy a large amount of space. Their power output is relatively low. Experiments are now going on to find a more efficient source of power. An obvious choice is nuclear power, and this is already being used in the U.S. Navy submersible *NR-1.* But *NR-1* cost $99.2 million— a price far beyond the reach of most manufacturers.

Industrial need for submersibles that can perform specific tasks has led to the development of a new type of undersea workboat. One such vessel is Perry Submarine's *Deep Diver,* which was designed by Edwin Link. *Deep Diver* can carry both working divers and non-diving observers and advisers. It consists of two compartments—a chamber that can be pressurized for the divers and a chamber kept at atmospheric pressure for the pilot and observers.

Because divers cannot at present work at depths below 1,000 feet in safety, mechanical aids, like the *DS-2000*'s manipulators, must be developed to help them. But it is difficult to carry out very complicated or heavy work with these devices. The average submersible, buoyant in the water, cannot exert sufficient pressure on a manipulator to make it effective. To overcome this difficulty, the vessel must be anchored in a way which will give it the necessary stability.

A British shipbuilding and engineering company, Cammell Laird, has devised an unusual submersible to meet this need. Their manned Sea Bed Vehicle is fitted with four giant wheels that enable it to travel across the sea floor, taking even steep slopes and undersea trenches in its stride. The Sea Bed Vehicle can carry heavy equipment to a depth of 600 feet and is fitted with manipulators controlled from inside by the crew. Its primary use is likely to be the setting up of underwater wellheads to obtain oil and gas from the North Sea.

The production of oil and gas offshore is already big business, in which submersibles and undersea habitats are becoming increas-

Below: an artist's impression of the Cammell Laird Sea Bed Vehicle. Operating at depths of up to 600 feet, it travels along the sea floor on huge hydraulically-driven wheels. Observations are made by means of a pair of T.V. cameras mounted on booms at the front. The vehicle provides a stable platform for heavy engineering work, such as setting up a wellhead. Divers use the rear, pressurized section—the front, at atmospheric pressure, houses pilot and observer.

ingly involved. By 1980, one third of the world's production of petroleum will come from under the ocean. Already much of the oil easily accessible in the shallow continental shelves has been found. As the search goes deeper the task of obtaining the oil by the use of equipment at the surface becomes impossible. In the future, men will need to live and work at an underwater drill site to produce the oil and patrol long pipelines leading back to shore.

There are other ocean riches beside oil. Diamonds, tin, iron, gold, phosphorus, and sulfur have been found in the continental shelves. All are now being commercially extracted by dredging from surface ships. Yet these minerals may well occur deeper than dredges can economically operate. Undersea mining camps may be set up to exploit them.

One last area holds promise of mineral wealth. Scientists think that the broad valley at the center of the giant mid-ocean-ridge system may contain an abundant store of heavy, metallic minerals. Miners living within this volcanic rift could harvest its riches.

The resources of the world's oceans are greater than those of the land. As world population grows and easily accessible land resources shrink, it is to the vast unexploited areas of the sea that man must turn for survival. Properly used, the ocean's riches could meet man's ever increasing demands for a very long time to come. Today, an all-out attempt is being made to locate and exploit those riches. But all man's efforts will come to nothing if he continues to pollute the sea as he is already doing.

Major disasters have focused world attention on the more obvious forms of pollution. In Europe, the tanker *Torrey Canyon* ran aground off southwest England and spilled much of her 117,000 tons of crude oil into the sea. The oil swept over 100 miles of England's lovely Cornish coast and was also carried across the English Channel to Brittany in France. The oil and the detergents used to remove it killed many thousands of seabirds, fish, and other sea creatures. Two years later, in 1969, about 230,000 gallons of oil poured from a leaking oil platform off the Santa Barbara Channel, California. Beaches were polluted and marine life was massacred along 20 miles of the Californian coastline.

Oil pollution kills thousands of sea creatures every year. But this is only a small part of the pollution problem. Already a flood of civilization's by-products has invaded the sea. In addition to oil, these pollutants include raw sewage, dangerous chemicals, radioactive wastes, unwanted explosives, tanks of poison gases left over from wartime, detergents, pesticides, insecticides—the list is, perhaps, more endless than the sea.

This steady flow of waste has begun to make its mark. And, in the Mediterranean, the problem has reached a point that some scientists term "biological suicide." In Italy, an estimated 70 per cent of beaches are polluted and 15 per cent totally poisoned. In Venice, polluted canals are helping to sink the city. And from Italy's holiday playgrounds come reports of dead and poisoned fish

Above: the Liberian oil tanker *Pacific Glory* on fire after a collision in the English Channel, in October, 1970. Some of her 72,000 tons of crude oil spilled into the sea and floated toward the beaches of southern England. Left: pollution of another kind—detergent foam on an English estuary. Below: the sad result of oil pollution in the sea—a dead gannet, its feathers clogged with oil, found washed up on a beach in southern Wales.

and algae, mutilated vegetation, and warnings that bathers must not use certain parts of the coast. According to the secretary of the Mediterranean symposium on marine pollution, "It is not a question as to *when* the sea will be dead. For the Italians it has already happened. The effect on other countries bordering the Mediterranean will only be a matter of time."

Such dire warnings are not confined to the Mediterranean. Pollution is a threat that knows no frontiers. Pollution of one part of one ocean will, sooner or later, pollute all parts of all oceans—from top to bottom. The churning, worldwide system of ocean currents guarantees this.

Already fish caught in many different parts of the world have been found to contain toxic substances. And these include deep-sea fish. Lead from gasoline has been discovered in Arctic snow. The ethnologist Thor Heyerdahl, crossing the Atlantic in his papyrus boat, found huge patches of oil even in the middle of that ocean. And these are but a few examples.

Statistics indicate that pollution is now growing three times faster than world population. In December, 1970, at the close of a conference on marine pollution, 415 scientists from all over the world urged an immediate survey of pollution in the oceans and called for a worldwide monitoring system to control it. In promoting an International Decade of Ocean Exploration for the 1970's, the United States has suggested that special emphasis be placed on "goals of protecting the oceans from the harmful effects of pollution." At the request of Sweden, the United Nations has agreed to organize a conference in Stockholm in 1972 to consider the whole question of man and his environment.

These moves highlight the international nature of the problem and the necessity for international solutions. What is needed is an even greater awareness of the dangers, coupled with a willingness to devote far more organization, equipment, knowledge, and money toward the prevention and control of pollution.

Scientists point out that, if the sea is carefully used, it can be both an important source of food and minerals *and* a safe disposal area for some wastes—particularly those that can be broken down into harmless substances by natural processes. The sea, they say, is man's last storehouse of resources on earth. He cannot afford to upset its life cycles, poison, or pollute it.

The awareness of the sea's importance for man's survival has drawn the nations of the world together in an unprecedented program of oceanic research. The 1970's mark the beginning of a new era in the exploration of the ocean. Now, as never before, man has the means to discover the sea's secrets and to reap its riches. It is up to him to use that ability wisely for the benefit of all mankind.

Right: effluent pours into Lake Erie, U.S.A., destroying the animal and plant life in the water. Where sewage and chemical effluents are pumped into the sea in uncontrolled amounts the balance of nature is similarly disturbed. Only coordinated international action will be able to save the life in the sea.

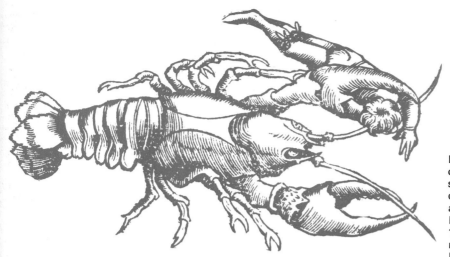

Left: "This is a shrimp or lobster of truly astonishing size and it is so powerful that with its forelegs it can seize a man swimming in the water and kill him." This is how Sebastian Munster, a German geographer of the 1500's, described one of the "horrible monsters of the northern seas" in his book *Cosmographia Universalis*.

Appendix

Ever since the first diver ventured a few feet beneath the sea, man has dreamed of conquering the deep. Today it seems that that dream may be realized. Undersea vessels have taken men to the deepest parts of the ocean. Divers roam freely in its upper levels and men have lived and worked beneath the waves. But the sea is slow to yield up its secrets. Even today, only a tiny part of the underwater world has been explored. The pioneers have opened the way to the depths. For those who follow them, all the excitement of high adventure and the thrill of discovery remain.

Now, as man embarks upon a new era of undersea exploration, he is continuing the work of generations of divers, seafarers, and scientists before him. Pitting himself against a hostile and largely unknown environment, the modern undersea explorer feels the same curiosity and fascination that set his predecessors on the long journey into inner space. It is this deep-rooted fascination for the sea that has been man's constant inspiration in his quest for greater understanding of the world beneath the ocean.

Secrets of the Sea has traced the course of that quest down the centuries to the present day. Here, in the appendix, the firsthand accounts of divers and contemporary records of their achievements bring to life in vivid terms the gradual developments of man's relationship with the sea. This is the human story behind the conquest of the deep, from the first timid penetration of the sea, through the battle against fear and danger, to the day when five young aquanauts could live in comparative comfort on the ocean floor.

For quick and easy reference to the explorers mentioned in this book, the appendix also contains a section of short individual biographies, set out in alphabetical order. Also included is a table outlining the principal stages which have marked the conquest of the depths. A glossary gives a fuller explanation of specialized terms and phrases used in the book, and this is followed by picture credit information and index.

Left: a diver wearing a helmet of the type used in the U.S. Navy's *Sealab* experiment. It was developed by the Kirby-Morgan Corporation, U.S.A., and contains an underwater communication unit. As divers probe deeper into the oceans they face problems similar to those of astronauts in outer space. Today's travelers in inner space are the first of a new generation of intrepid aquanauts.

The Perils of Sponge Diving

The sponge divers of ancient times often paid with their lives for braving the pressures of the deep. But more than the sea itself, they feared the deadly monsters said to live there. Here the Greek poet Oppian describes the divers' ordeals.

"As soon as he has severed the sponge from the rock he pulls the cord to let his companions above know that he must now be pulled to the surface as quickly as possible. Once cut away from the rock, the sponge bleeds a nauseating liquid which spreads around the diver and is sometimes sufficient to kill him, so offensive is the smell to the nostrils of man. . . .

"Sometimes his efforts end miserably and cruelly. Once he has plunged into the waves the unfortunate often does not return. He has encountered some hidden monster of the deep and lost his life. Desperately at first he pulls the rope in order to be hoisted to safety, but the monster seizes him and then a horrible tug of war takes place. The monster holds him from below and his friends pull him from above, disputing the half-devoured corpse of the unfortunate diver between them. Then with heavy hearts his companions hasten away from the ill-fated area abandoning a hopeless undertaking."

Halieutics, III, *Oppian, from Man and the Underwater World by Pierre de Latil & Jean Rivoire, trans. by Edward Fitzgerald (Jarrolds Publishers (London) Limited: 1956) pp. 46–47.*

Below: an engraving of an African sponge diver. Like the divers of ancient Greece, he has no artificial air supply but relies on the air in his lungs as he cuts sponges from the seabed. When he signals, his friends pull him to the surface.

Monsters of the Deep

Above: "The horrible monsters of the northern seas" as depicted by Sebastian Munster in *Cosmographia Universalis*. Amidst his monsters, though, he has drawn a "mercy fish" that "prevents drowned sailormen from being devoured by other sea monsters."

Stories of huge and hideous sea monsters continued to intimidate sailors for many hundreds of years. In this illustration (left) of 1544, the German geographer Sebastian Munster depicted some of the terrifying beasts thought to lurk in wait for the Atlantic seafarer. They are described by him below.

"Giant fish, such as whales, sea cows, and similar creatures, all as big as mountains, are to be seen off Iceland. Unless they are first frightened off by the sound of trumpets, or their attention is distracted by the casting of round, empty vessels into the sea for them to play with, they capsize the vessels they fall in with. It sometimes happens that the captains of ships fasten their anchors into such giant beasts, believing them to be islands, and then the vessels are in danger of being dragged under the surface. The inhabitants of Iceland call these beasts 'Trolual,' which, in their own language, means 'Devil Whales.' Even today many of the inhabitants of Iceland build their houses of the bones of such monsters. . . .

"There are many other kinds of monsters, both fish and bird, in Northern waters, the which, with their astonishing forms, could fill a whole book if one wished to disentangle the great variety of species with which God has populated the colder regions."

Cosmographia Universalis *Sebastian Munster from Man and the Underwater World by Pierre de Latil & Jean Rivoire, trans. by Edward Fitzgerald (Jarrolds Publishers (London) Limited: 1956) p. 48.*

Right: a sketch made by Captain Taylor of his barque *British Banner* being attacked by a sea serpent on April 25, 1860. He estimated the serpent to be about 300 feet long with a "black back, shaggy mane, horn on the forehead, and large glaring eyes . . ."

Cleopatra's Joke

Right: an engraving showing Cleopatra and Mark Antony on a fishing trip at Alexandria. Mark Antony used divers to plant fish on his line, but Cleopatra was not fooled and countered with a more telling version of the same joke.

Diving in ancient times was not always difficult and terrifying. It had its comic side, too, as this account by the Greek writer Plutarch shows. Plutarch tells how, during a fishing trip at Alexandria in Egypt, the Egyptian queen Cleopatra made use of divers to play a practical joke on her lover, the Roman general Mark Antony.

"He [Antony] went out one day to angle with Cleopatra, and, being so unfortunate as to catch nothing in the presence of his mistress, he gave secret orders to the fishermen to dive under water, and put fishes that had been already taken upon his hooks; and these he drew so fast that the Egyptian [Cleopatra] perceived it. But, feigning great admiration, she told everybody how dexterous Antony was, and invited them next day to come and see him again. So, when a number of them had come on board the fishing boats, as soon as he had let down his hook, one of her servants was beforehand with his divers, and fixed upon his hook a salted fish from Pontus [a region on the Black Sea]. Antony, feeling his line give, drew up the prey, and when, as may be imagined, great laughter ensued, 'Leave,' said Cleopatra, 'the fishing-rod, General, to us poor sovereigns of Pharos and Canopus; your game is cities, provinces and kingdoms.'"

Lives: *Plutarch, trans. by J. Dryden, ed. by A. H. Clough (Nimmo: London, 1893) p. 53.*

The Boy and the Dolphin

Dolphins are among the best-loved creatures of the sea. Some of the earliest legends tell of their affection for mankind in general and for children in particular. Here, the Roman writer Pliny describes a dolphin's friendship for a boy.

"In the daies of Augustus Caesar the Emperour, there was a Dolphin . . . which loved wonderous well a certain boy, a poore mans sonne: who using to go every day to schoole . . . was woont also about noone-tide to stay at the water side, and to call unto the Dolphin, *Simo, Simo,* and many times would give him fragments of bread. . . . Well, in processe of time, at what houre soever of the day, this boy lured for him and called *Simo* . . . out he would and come abroad, yea and skud amaine [swim swiftly] to this lad . . . and would gently offer him his backe to mount upon. . . . Thus when he had him once on his backe, he would carrie him over the broad arme of the sea . . . to schoole; and in like manner convey him backe againe home: and thus he continued for many yeeres. . . . But when the boy was falne sicke and dead, yet the Dolphin . . . came [regularly] to the woonted place, & missing the lad, seemed to be heavie and mourne again, untill for verie griefe and sorrow . . . he also was found dead upon the shore."

Naturall Historie of G. Plinius Secundus, Vol. I, *trans. by Philemon Holland (London : 1601) p. 239.*

Above: an engraving of Arion on a dolphin. Since the earliest legends, tales of the sea have always remarked the dolphin's affection for mankind.

Below: two common dolphins leap from the sea in perfect formation. These graceful sea mammals frequently swim alongside ships, much to the delight of sailors, who regard them as lucky.

The Underwater Ark

Above: English bishop, John Wilkins. He was passionately interested in the sea and, in 1648, proposed a means by which the human race would be able to adapt itself to life under the water.

As early as 1648, the sea-loving English bishop John Wilkins was predicting the age of underwater man. In the following extract from his writings, Wilkins points out the advantages of an "Ark for Submarine Navigations."

"1. 'Tis *private*; a man may thus go to any coast of the world invisibly, without being discovered or prevented in his journey.

"2. 'Tis *safe*; from the uncertainty of *Tides*, and the violence of *Tempests* . . . from *Pirates* and *Robbers* . . . from ice and great frosts which doe so much endanger the passage towards the Poles.

"3. It may be of very great advantage against a Navy of enemies, who by this means may be undermined in the water and blown up.

"4. It may be of a special use for the relief of any place that is besieged by water, to convey unto them invisible supplies . . . .

"5. It may be of unspeakable benefit for submarine experiments and discoveries: as,

"The deep caverns and subterraneous passages where the sea-water in the course of its circulation doth vent itself into other places and the like. The nature and kinds of fishes, the severall arts of catching them, by alluring them with lights, by placing nets around the sides of this Vessell, shooting the greater sort of them with guns. . . . These fish may serve not only for food, but for fewell likewise, in respect of that oyl which may be extracted from them.

"The many fresh springs that may probably be met with in the bottom of the sea will serve for the supply of drink and other occasions.

"But above all, the discovery of submarine treasures is more especially considerable, not only in regard of what hath been drowned by wrecks, but the several precious things that grow there, as Pearl, Coral Mines, with innumerable other things of great value. . . .

"All kinds of arts and manufactures may be exercised in this Vessell. The observations made by it, may be both written and (if need were) printed here likewise. Severall Colonies may thus inhabit, having their children born and bred up without the knowledge of land, who could not chuse but be amazed . . . upon the discovery of this upper world."

Mathematical Magick, Vol. II, *John Wilkins (London: 1648) pp. 4–5.*

The First Submarine Attack

The first submarine to be used against an enemy warship was David Bushnell's hand-operated *Turtle*. In August, 1776, during the Revolutionary War, the *Turtle* made an unsuccessful attack on the British vessel H.M.S. *Eagle* in New York harbor. Here Bushnell describes the attack in a letter of 1787 addressed to Thomas Jefferson, who was then American minister in Paris.

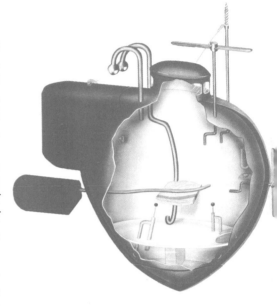

"I sent one (Sgt. Ezra Lee) . . . from New York to a fifty-gun ship lying not far from Governor's Island. He went under the ship and attempted to fix the Woodscrew [attached to a 150-pound charge of gunpowder] into her bottom, but struck, as he supposes, a bar of iron. . . . Not being skilled in the management of the Vessel, in attempting to move to another place, he lost the Ship . . . .

"In his return from the Ship to N. York, he passed near Governor's Island, and thought he was discovered by the enemy, on the Island; being in haste to avoid the danger he feared, he cast off the magazine [containing the gunpowder] as he feared it retarded him . . . . After the magazine had been cast off, one hour, the time the internal apparatus was set to run, it blew up with great violence."

Captain Cousteau's Underwater Treasury *ed. by J-Y. Cousteau and James Dugan (Hamish Hamilton Ltd.: London, 1960) p. 121.*

Above: an artist's impression of the *Turtle*, a one-man submarine built by David Bushnell for military use.

Below: the British fleet leaving Boston harbor, in April 1776, during the Revolutionary War. In August, the *Turtle* unsuccessfully attacked one of the British ships in New York harbor.

Beneath the North Pole

On August 3, 1958, the U.S. Navy's nuclear submarine *Nautilus* made history by sailing under the North Pole. Threading her way beneath the Arctic ice, the *Nautilus* pioneered a Northwest Passage from the Pacific to the Atlantic. The final stages of her epic voyage are recorded here by a member of her crew, Lieutenant William G. Lalor Jr.

"July 30—1 A.M. [In the Chukchi Sea]. . . . Sighted our first ice of the trip. . . . It was a lone floe, about 100 yards long, which we easily dodged. It looked like a beautiful sailing ship moving majestically by, reflecting a rainbow of colors.

"Cruising slowly south, we began a disheartening routine that held for the next twenty-four hours. Since ice loomed to the west and the sea that way was shallow, we turned east and then north again, hoping that the pack boundary would be closer to deep water along the new approach. Our object was to reach at least 300 feet of ice-free water before diving. . . .

"At 4.37 A.M. on August 1, we cruised north of Point Barrow, invisible just over the horizon. The Fathometer, whose moving arm had been monotonously showing 160 to 180 feet of water now suddenly indicated 420 feet. There was jubilation in the voice of Chief of the Watch John J. Krawczyk as he called the bridge on the intercom. We were there—in a tongue of the deep Barrow Submarine Canyon, which should lead us north to the even deeper Arctic Ocean. . . .

"The diving-alarm honked twice; in a minute *Nautilus* slipped beneath the sea. The ship eased down to 200 feet, 300, then deeper, as we followed the ever-deepening bottom. Hereafter, we would be measuring depth of water in hundreds of fathoms rather than feet. . . .

"August 2. At latitude 76° 22', soundings went from about 2,000 fathoms to 500 fathoms very abruptly. We were crossing a 9,000-foot submerged mountain range, uncharted and unknown. This feature continued for 70 miles. . . .

"80° N. 600 miles from the Pole, and 1,200 miles from the ice edge in the Atlantic. . . . Passing through fairly light ice now . . . as we near the Pole of Inaccessibility. This imaginary area is supposed to be the hardest to reach in the Arctic, almost the geographic center of the ice pack. . . .

"August 3—10 A.M. Latitude 87° N. Passing history's and our

Above: the U.S. Navy nuclear submarine *Nautilus* leaving for surface trials off Long Island Sound in January, 1955.

Left: the conning tower of the American nuclear submarine *Sargo* breaks through the ice at the North Pole on March 8, 1960. It cruised 2,744 miles under the ice to become the third American submarine to reach the pole.

Right: pack ice. Pack ice can be as much as 65 feet deep in places where there are ridges and hummocks.

Below: *Nautilus* docking at Portland, England, on August 12, 1958. It had just completed a historic journey from the Pacific to the Atlantic, through the waters below the North Pole.

farthest point north by ship.... Dinner was delayed to allow party preparations to go on in the crew's mess. Leading Cook Jack L. Baird put the finishing touches on his North Pole cake, with the replica of our polar flag as icing.... I sat with the captain in the wardroom as he signed letters and put the finishing touches on those to the President and to the ship's sponsor, Mrs. Eisenhower. Frank Adams came in.

" 'Two miles to go, Captain.'

"The captain spoke briefly and movingly over the intercom:

" 'With continued good fortune, *Nautilus* will soon accomplish two goals long sought by those who sail the seas. First, the opening of a route for rapid voyages between the great Pacific and Atlantic oceans. Second, the attainment of the North Pole by ship.

" 'Thus our remarkable ship has been blessed with her greatest opportunity—the discovery of the only truly practicable Northwest Passage. On this historic Sunday, August 3, 1958, let us offer our

thanks to Him who has blessed us with this opportunity and who has guided us so truly....'

"We observed a moment of silent prayer. As *Nautilus* approached the Pole, the captain began a countdown: '8 ... 6 ... 4 ... 2 ... 1 ... Mark! August 3, 1958. Time, 2315 (11.15 P.M.) Eastern Daylight Saving Time for the United States and the United States Navy, the North Pole.'

"A dream had become reality. We had arrived."

Submarine Through the North Pole *Lt. William G. Lalor Jr., National Geographic Magazine © 1959 by the National Geographic Society, Washington, D.C.*

Deaf and Dumb Beneath the Waves

The development of the helmet suit was a major breakthrough for diving. But, inside the device, the diver had little freedom of movement and was dependent on air pumped down to him. The Scottish novelist Robert Louis Stevenson tried out a hard hat suit in the 1860's and here describes the experience.

"It was grey, harsh, easterly weather [and] the swell ran pretty high . . . when I found myself at last on the divers' platform, twenty pounds of lead upon each foot, and my whole person swollen with play and ply of woollen underclothing. One moment, the salt wind was whistling round my nightcapped head; the next, I was crushed almost double under the weight of the helmet. As that intolerable burthen was laid upon me, I could have found it in my heart (only for shame's sake) to cry off the whole enterprise. But it was too late. The attendants began to turn the hurdy-gurdy, and the air to whistle through the tube; someone screwed in the barred window of the vizor; and I was cut off in a moment from my fellow-

Above: divers preparing for work in 1873. Diving then was an exhausting business, a pastime only to be undertaken by strong and determined men.

Left: a photograph taken in 1891 shows a diver wearing a suit designed for work at a depth of 30 feet.

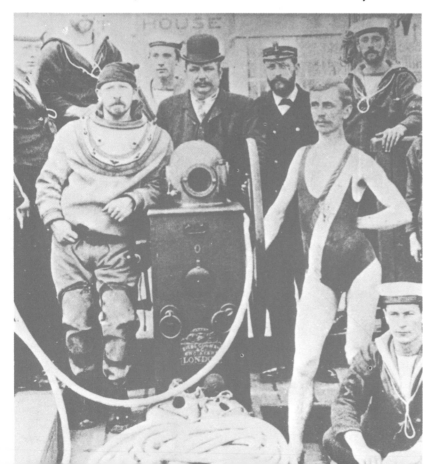

Right: a diver in full equipment in 1870. Looking and feeling like men in suits of armor, divers descended to a silent world beneath the waves. The silence was broken only by the noise of their swallowing and by the hiss of air inside their helmets.

men; standing there in their midst . . . a creature deaf and dumb, pathetically looking forth upon them from a climate of his own. . . . But time was scarce given me to realize my isolation; the weights were hung upon my back and breast, the signal-rope was thrust into my unresisting hand; and setting a twenty-pound foot upon the ladder, I began ponderously to descend.

"Some twenty rounds below the platform twilight fell. Looking up, I saw a low green heaven mottled with vanishing bells of white; looking around . . . nothing but a green gloaming, somewhat opaque but very restful and delicious. Thirty rounds lower . . . a dumb helmeted figure took me by the hand, and made a gesture (as I read it) of encouragement; and looking in at the creature's window, I beheld the face of Bain [an engineer-diver]. There we were, hand to hand and (when it pleased us) eye to eye and either might have burst himself with shouting, and not a whisper came to his companion's hearing. . . .

"How a man's weight, so far from being an encumbrance, is the very ground of his agility, was the chief lesson of my submarine experience. The knowledge came upon me by degrees. As I began to go forward with the hand of my estranged companion . . . Bob [Bain] motioned me to leap upon a stone; I looked to see if he were possibly in earnest; and he only signed to me the more imperiously. Now the block stood six feet high; it would have been quite a leap for me unencumbered; with the breast- and back-weights, and the twenty pounds upon each foot, and the staggering load of the helmet, the thing was out of reason. I laughed aloud in my tomb; and to prove to Bob how far he was astray, I gave a little impulse from my toes. Up I soared like a bird, my companion soaring at my side. As high as the stone, and then higher. . . . Even when the strong arm of Bob had checked my shoulders, my heels continued their ascent; so that I blew out sideways like an autumn leaf, and must be hauled in, hand over hand, as sailors haul in the slack of a rope, and propped upon my feet like an intoxicated sparrow. . . .

"It is bitter to return to infancy, to be supported, and directed and perpetually set upon your feet, by the hand of someone else. The air besides, as it is supplied to you by the busy millers on the platform, closes the Eustachian tube [passage through which air passes from the throat to the middle ear] and keeps [you] perpetually swallowing, till the throat is grown so dry that [you] can swallow no longer. And for all these reasons—although I had a fine, dizzy, muddleheaded joy in my surroundings, and longed, and tried, and always failed to lay my hands on the fish that darted here and there about me, swift as humming-birds—yet I fancy I was rather relieved than otherwise when Bain brought me back to the ladder and signed to me to mount."

Random Memories *Robert Louis Stevenson from The Works of Robert Louis Stevenson, Vol. XVI (Chatto and Windus: London, 1912) pp. 170–173.*

Attack by Sharks!

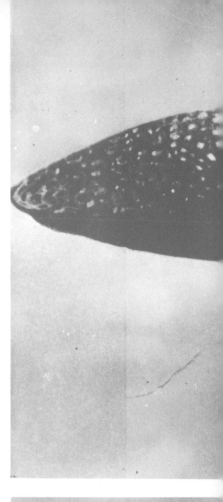

Free divers have learned a great deal about the behavior of some of the most dangerous undersea creatures. According to Jacques-Yves Cousteau, "One can never tell what a shark is going to do." In this gripping account of a meeting with hostile sharks, Cousteau proves his point.

"A few lengths down the line in a depth of fifteen feet, we sighted an eight-foot shark of a species we had never before seen. He was impressively neat, light gray, sleek, a real collector's item. A ten-inch fish with vertical black-and-white stripes accompanied him a few inches above his back, one of the famous pilot fish. We boldly swam toward the shark, confident that he would run as all the others had. He did not retreat. We drew within ten feet of him, and saw all around the shark an escort of tiny striped pilots. . . . Sea legends hold that the shark has poor eyesight and pilot fish guide him to the prey, in order to take crumbs from his table. . . .

"The handsome gray was not apprehensive. I was happy to have such an opportunity to film a shark, although, as the first wonder passed, a sense of danger came to our hearts. Shark and company slowly circled us. . . . The shark made no hostile move nor did he flee, but his hard little eyes were on us. . . .

"The shark had gradually led us down to sixty feet. Dumas pointed down. From the visibility limit of the abyss, two more sharks climbed toward us. They were fifteen-footers, slender, steel-blue animals with a more savage appearance. . . .

"Our old friend, the gray shark, was getting closer to us, tightening his slowly revolving cordon. . . . We revolved inside the ring, watching the gray, and tried to keep the blues located at the same time. We never found them in the same place twice. . . .

"Dumas and I ransacked our memories for advice on how to frighten off sharks. *'Gesticulate wildly,' said a lifeguard.* We flailed our arms. The gray did not falter. *'Give 'em a flood of bubbles,' said a helmet-diver.* Dumas waited until the shark had reached his nearest point and released a heavy exhalation. The shark did not react. *'Shout as loud as you can,' said Hans Hass.* We hooted until our voices cracked. The shark appeared deaf. *'Cupric acetate tablets fastened to leg and belt will keep sharks away if you go into the drink,' said an Air Force briefing officer.* Our friend swam through the copper-stained water without a wink. His cold, tranquil eye appraised us. He seemed

Left: a diver about to kill a
nurse shark with an explosive device.
Nurse sharks are usually inoffensive
but they will bite when annoyed.

Above: diver George Mayer plays tag
with a whale shark. He is in little
danger of being bitten, but he keeps
well clear of its powerful tail.

to know what he wanted, and he was in no hurry.

"A small dreadful thing occurred. The tiny pilot fish on the shark's snout tumbled off his station and wriggled to Dumas. It was a long journey for the little fellow, quite long enough for us to speculate on his purpose. . . .

"Instinctively I felt my comrade move close to me, and I saw his hand clutching his belt knife. Beyond the camera and the knife, the gray shark retreated some distance, turned and glided at us head-on.

We did not believe in knifing sharks, but the final moment had come, when knife and camera were all we had. I had my hand on the camera button and it was running, without my knowledge that I was filming the oncoming beast. The flat snout grew larger and there was only the head. I was flooded with anger. With all my strength I thrust the camera and banged his muzzle. I felt the wash of a heavy body flashing past and the shark was twelve feet away, circling us as slowly as before, unharmed and expressionless. . . .

"The blue sharks now climbed up and joined us. Dumas and I decided to take a chance on the surface. We swam up and thrust our

173

Attack by Sharks!

masks out of the water. The *Élie Monnier* [Cousteau's surface vessel] was three hundred yards away, under the wind. We waved wildly and saw no reply from the ship. We believed that floating on the surface with one's head out of the water is the classic method of being eaten away. Hanging there, one's legs could be plucked like bananas. I looked down. The three sharks were rising towards us in a concerted attack.

"We dived and faced them. The sharks resumed the circling maneuver. As long as we were a fathom or two down, they hesitated to approach. . . . We evolved a system of taking turns for brief appeals on the surface, while the low man pulled his knees up against his chest and watched the sharks. A blue closed in on Dumas's feet while he was above. I yelled. Dumas turned over and resolutely faced the shark. The beast broke off and went back to the circle. . . . We saw no evidence that our shipmates had spied us.

"We were nearing exhaustion, and cold was claiming the outer layers of our bodies. I reckoned we had been down over a half hour. Any moment we expected the constriction of air in our mouth-pieces, a sign that the air supply is nearing exhaustion. . . . There was five minutes' worth of air in the emergency ration. When that was gone, we could abandon our mouthpieces and make mask dives, holding our breath. That would quicken the pace, redouble the drain on our strength, and leave us facing tireless, indestructible creatures that never needed breath.

"The movements of the sharks grew agitated. They ran around us, working all their strong propulsive fins, turned down and disappeared. We could not believe it. Dumas and I stared at each other. A shadow fell across us. We looked up and saw the hull of the *Élie Monnier*'s launch. Our mates had seen our signals and had located our bubbles. The sharks ran when they saw the launch."

The Silent World *J-Y. Cousteau with Frédéric Dumas (Hamish Hamilton: London, 1953) pp. 120–123.*

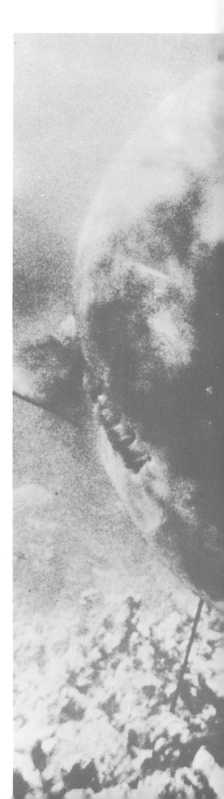

Right: diver Ron Zangari harpoons a 12-foot tiger shark at Swain Reefs off Australia. At a depth of about 120 feet the blood pumping from a wounded animal appears bright green to a diver. This is because the red light in the sun's rays only penetrates a few feet down into the sea.

Where Blood is Green

The twilight zone, about 120 feet down in the sea, is filled with dazzling colors that can only be seen with artificial light. To the diver, this zone looks blue and green. With increased depth, colors change and are gradually lost, as the sun's rays are selectively filtered by the sea water. The startling effect that these changes can produce is described here by Jacques-Yves Cousteau.

"Once we were hunting under the isolated rocks of La Cassidaigne [in the Mediterranean]. Twenty fathoms down Didi [Dumas] shot an eighty-pound liche [a tuna-like fish]. The harpoon entered behind the head but missed the spinal column. The animal was well hooked but full of fight. It swam away, towing Didi on his thirty-foot line. When it went down, he held himself crosswise to cause a drag. When the fish climbed, Didi streamlined himself behind it and kicked his flippers to encourage ascent.

"The liche fought on untiringly. It was still struggling as we drew near the depletion of our air supply. Dumas hauled himself forward on the line. The fish circled him at a fast, wobbling speed, and Dumas spun with it to avoid being wound up in the line. . . . Dumas hauled in the last feet of cord, and got a grip on the harpoon shaft. He flashed his belt dagger and plunged it into the heart of the big fish. A thick puff of blood stained the water.

"*The blood was green.* Stupefied by the sight, I swam close and stared at the mortal stream pumping from the heart. It was the color of emeralds. Dumas and I looked at each other wildly. We had swum among the great liches as aqualungers and taken them on goggle-dives, but we never knew there was a type with green blood. Flourishing his astounding trophy on the harpoon, Didi led the way to the surface. At fifty-five feet the blood turned dark brown. At twenty feet it was pink. On the surface it flowed red.

"Once I cut my hand seriously one hundred and fifty feet down and saw my own blood flow green. I was already feeling a slight attack of rapture. To my half-hallucinated brain green blood seemed like a clever trick of the sea. I thought of the liche and managed to convince myself that my blood was really red."

The Silent World *J-Y. Cousteau with Frédéric Dumas (Hamish Hamilton: London, 1953) pp. 140–141.*

175

Return from the Deep

Robert Sténuit was the first man to live for 26 hours at a depth of 200 feet. The Belgian diver was to spend two days in the Mediterranean testing Edwin Link's 3-foot by 11-foot aluminum cylinder, but the arrival of the violent mistral wind cut short the experiment. As the cylinder pitched and rolled in a stormy sea, Sténuit made his difficult ascent.

"I start the winch, *up*: nothing happens; *down*: nothing happens. Another try: still nothing. The next second I hear a long, long scraping. All the slack of the chain is freely sliding through the spinning rack. . . . The axis pin of the rack has given way under the jolts of the short nasty mistral surge, and I am floating freely, in mid-water. . . . I can see Sullivan, one of our divers, through a porthole struggling with the chain. He tries to tie a line on it, to get hold of the cylinder, I guess. But no sooner has he caught it than it slips through his fingers like an hysterical snake. . . . He surfaces with me, clinging to the cylinder.

Right: Robert Sténuit and Jon Lindbergh re-enact their exit from the decompression chamber in which they had lived for 49 hours at a depth of 432 feet. This dive, in July, 1964, set a new deep-diving record.

Below: Belgian diver Robert Sténuit (left) and Edwin Link inspect the diving cylinder. In September, 1962, Sténuit spent 26 hours in this 11-foot-long aluminum cylinder 200 feet underwater in the Mediterranean Sea.

"It is not the decompression that worries me. As long as my internal pressure is maintained, it matters little where the cylinder may be wandering, but what could be serious is a head-on crash of the cylinder on the hull of *Sea Diver* [Link's yacht]. . . .

"But not at all; the cylinder breaks through the surface far from the yacht. It is night. The sea seems frightfully rough. . . . I am tossed around in my cylinder like dice in a dice box. Through the porthole I see the black sky, then the white ship, then the blue water—all dancing a mad ballet. Chains are lashing everywhere.

My books and food drop in disorder at the bottom of the cylinder, in a well of foaming water. . . .

"I am concerned for the diver who has now to shackle the second chain to the cylinder, the chain that makes it possible to haul me, horizontally, on board. A backlash of it could tear out one of his arms, and I prefer not to imagine him caught between the cylinder and the hull. . . .

"A quarter of an hour has passed, half an hour; I am still bouncing on the surface in the oblique, jolting cylinder. . . .

" 'There you are, you are going up now.' My back to the wall and pushing out with feet and arms, I try hard to hold fast in my cock-tail shaker. The water which fills the bottom of the lock splashes everywhere and I am drenched. Bang! A heavy blow on *Sea Diver*. . . . I can hear Ed [Link] shouting orders . . . Zzing! Another hammering ram smashes the awning. . . . More overpowering blows. I can't do anything but hang on teeth and nails, and wait. A final jolt, and [a] relieved voice . . . 'You have arrived. The cylinder lies in its cradle. All is well.' "

The Deepest Days *Robert Sténuit, trans. by Morris Kemp (Hodder and Stoughton Limited: London, 1967) pp. 97–99.*

Above: a diagram showing how Link's diving cylinder (left) conveyed divers to the SPID (Submersible Portable Inflatable Dwelling). Umbilical cables from the surface supply the cylinder with breathing gases, electricity, and a telephone link.

Undersea Vacation

On July 6, 1970, five American women began a "summer vacation" on the ocean floor. Sylvia Earle, Renate True, Ann Hartline, Alina Szmant and Peggy Lucas were taking part in the *Tektite II* underwater habitat program. The five women spent two weeks under the sea. This is how Peggy Lucas, engineer on the project, describes life in an underwater home.

"There was never any claustrophobia, and, as Ann Hartline put it, 'we got along unbelievably well.' The only times we disagreed came when we all wanted to go out swimming at the same time.

"We solved these conflicts by the time-honored practice of flipping a coin. Usually, of course, I remained "inside" with my engineering duties. The other aquanettes generally swam in pairs, using the 'buddy system' so that one of them could always be on the alert for possible danger while the other concentrated on her research. Around the habitat were bubble-domed steel cages into which the girls could swim if sharks were in the area. The cages even had telephones which permitted swimmers to talk with me in the habitat or communicate with the support crews on the surface. . . .

"We loved to watch the various forms of sea life as they made their way past our habitat. Our 'pet' was a four-foot barracuda that inhabited the entrance-way but gave us no problems. He seemed to enjoy posing for pictures—he even bared his teeth for the camera. . . .

"Our daily diet consisted of frozen foods most of which we heated and ate from plastic dishes. Our menus included soups, meats, stews, rolls, butter, vegetables, and, yes, ice cream. . . .

"We used a sort of 'dumb waiter' to deliver and receive mail from upstairs. The same contrivance also took away the used plastic dishes, spoons and forks, and garbage, and brought us supplies. . . .

"We had no medical, physiological or psychological problems. As Sylvia Earle said, 'It was not a hardship; it was a genuine pleasure' . . . Actually, we all were a bit sad to leave the habitat on July 20. . . . There was just too much to see and learn in only 14 days. All of us would like to go back again if given the opportunity. It was just beautiful!

"We didn't really feel all that 'alone' in the deep. . . . During all the time we were on the ocean floor, we could hear the engine of the safety boat on the surface, and we knew that there were divers aboard, ready to come to our aid in an emergency. . . .

"Each of us has been asked if, having gone down to the bottom of the sea as an aquanaut, she would like to go into space as an astronaut. Alina spoke for all of us, I think, when she said: 'That's not our field. But if they discover oceans on the moon, I think I'd like to go.' "

Two Weeks in a Cottage on the Bottom of the Sea *Peggy Lucas* (*Science Horizons No. 117: October, 1970*).

Above: a woman aquanaut or "aquanette" returning to *Tektite II*, a habitat resting on the bottom 50 feet below the surface of the Caribbean Sea. In July, 1970, five women stayed for two weeks in this underwater home.

Left: a scene outside the habitat, with several members of the team at work. During their stay they studied marine life in the surrounding sea.

Right: in the habitat control room, Peggy Lucas (left) chats with team leader Sylvia Earle. As the team's engineer, Peggy had to stay inside the habitat for most of the time.

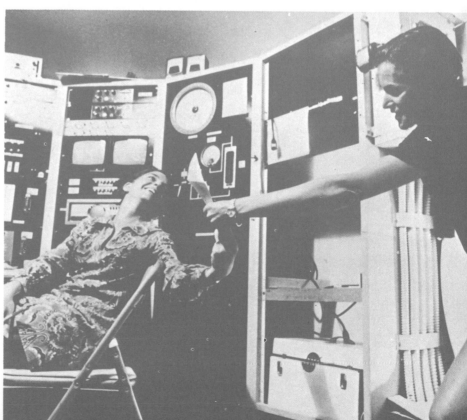

Probing the Depths

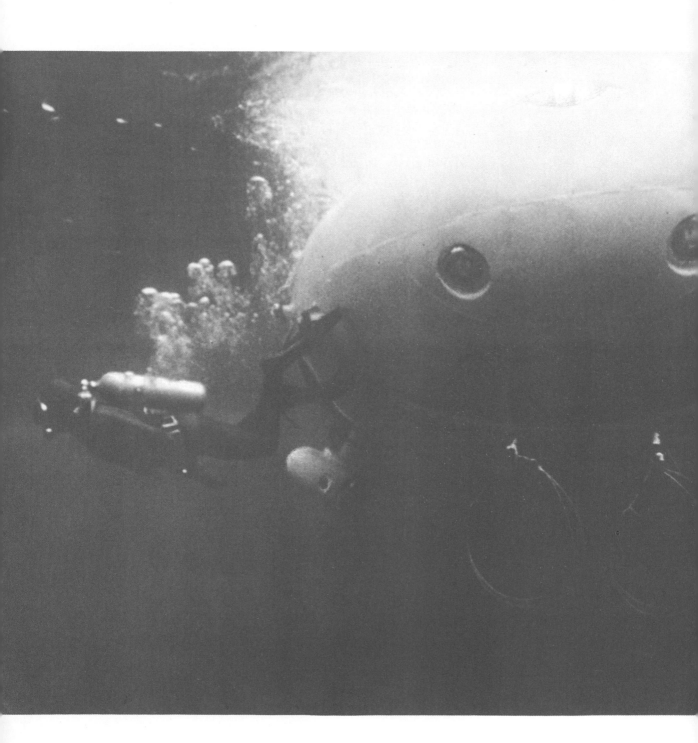

Right: this table shows the most significant diving achievements from 1934 to 1970, arranged in order of depths reached. The dates given in the right-hand column indicate actual achievements, but the design depths of several submersibles are also shown. Below: a typical free-swimming submersible, *Deep Quest*. This American submersible is designed to operate at a depth of 8,000 feet. It can take a crew of 4 and remain submerged for 24 hours at a time.

Depth (in feet)	Diver, vessel, or habitat	Date
33	Cousteau's *Conshelf I* steel cylinder habitat. Two divers stayed in it for one week.	1962
36	Cousteau's *Conshelf II* habitat. Five interconnected cylinders assembled in a star-like pattern. Five divers for one month.	1963
50	*Tektite II* twin-cylinder habitat. Series of divers for 2-week-20-day periods. U.S.A.	1970
82	*Sadko-2,* U.S.S.R. twin-sphered habitat. Six days.	1967
90	Advance cylinder of *Conshelf II.* Two divers for one week.	1963
100	*Tektite II* "minitat." Teams of two for 2-week or 20-day periods.	1970
192	*Sealab I,* steel capsule habitat. Four U.S. Navy divers for 10 days.	1964
200	*Man in Sea.* Link aluminium cylinder tested by Robert Sténuit for 26 hours. U.S.A.	1962
206	*Sealab II,* steel cylinder habitat. Three teams of 10 men for 15- and 30-day periods.	1965
330	*Conshelf III,* two-story steel sphere. Six divers for 22 days.	1965
432	*Man in Sea.* Link's SPID (Submersible Portable Inflatable Dwelling). Divers Robert Sténuit and Jon Lindbergh for 49 hours.	1964
547	Helmet dive by William Bollard (R.N. diver).	1948
610	*Scalab III,* steel cylinder habitat.	
650	*Man in Sea* experimental pressure chamber in laboratory. 48 hours.	1965
1,000	"Free dive" with independent breathing device. Hannes Keller and Peter Small. Small died.	1962
1,706	Experimental dive in laboratory. France.	1970
2,000	*Shinkai* submersible, Japan.	
2,500	*Auguste Piccard* submersible, Switzerland.	
3,028	*Bathysphere.* William Beebe and Otis Barton. U.S.A.	1934
4,462	*Benthoscope.* Otis Barton. U.S.A.	1949
5,000	*Vickers Pisces* submersible, Canada.	
7,100	*Sever II* submersible, U.S.S.R.	
15,000	*Aluminaut* submersible, U.S.A.	
20,000	*Deepstar 20,000* submersible, U.S.A.	
20,000/36,000	*Trieste II* (Interchangeable pressure hulls), U.S.A.	
35,800	Bathyscaph *Trieste.* Jacques Piccard and Lt. Don Walsh. Challenger Deep, Mariana Trench, southwest of Guam in the western Pacific. U.S.A.	1960
36,000	*Archimède,* France.	

Explorers of the Sea

AGASSIZ, ALEXANDER
1835–1910　　　　　　Switzerland
1872–1876: Sailed as a zoologist on
the *Challenger* expedition, making a
particular study of sea urchins.
1877–1880: Headed an American
scientific expedition aboard the U.S.S.
Blake. Explored the Caribbean Sea, the
Gulf of Mexico, and the Florida coast.
Was the first to use steel cables for
deep-sea dredging. Invented a new
type of trawl and a tow net for collecting
animal specimens from different depths.
1888–1902: Led a team of scientists
on a voyage to the Pacific Ocean
aboard the research vessel U.S.S.
Albatross. Made extensive soundings
and studies of marine life from Easter
Island and Callao, Peru, to the Bering
Sea. Also studied the waters off Japan
and the Sea of Okhotsk, Russia.
See Challenger *map on page 40*

ALBERT I, PRINCE OF MONACO
1848–1922　　　　　　　　Monaco
1885–1888: Went on a series of
expeditions to the Atlantic aboard his
yacht *Hirondelle*. Took numerous
soundings and temperature measurements throughout the Atlantic and
studied North Atlantic currents.
1892–1907: Sailed on 18 voyages in
his yachts *Princess Alice* and *Princess
Alice II*. Made wide-ranging oceanographic studies in the Mediterranean
and the Atlantic.
1910: Founded the Oceanographic
Museum in Monaco.
1911–1915: Continued to carry out
further marine research from his yacht
the *Hirondelle II*. Published many books
about his voyages, and encouraged
interest in oceanography. Became
known as the *Scientist Prince*.

BARTON, OTIS
1900's　　　　　　　　United States
1930: In the first bathysphere, a
machine for deep-sea exploration of his
own design, descended with William
Beebe to a depth of 800 feet.
1930–1934: Accompanied Beebe on
more than 30 other descents.
1948: Reached over 4,000 feet in the
benthoscope, which he had designed.
See entry under Beebe

BEEBE, WILLIAM
1877–1962　　　　　　United States
1899–1929: Led scientific expeditions
to South America, Borneo, and
Trinidad. Made hundreds of helmet
dives to study underwater life.
1930: Descended with Otis Barton in
the bathysphere, when they became
the first to reach a depth of 800 feet.
1930–1934: With Barton, made more
than 30 descents in the bathysphere.
One of the first set a record of 1,426 feet
and dive number 32 broke all previous
records by reaching a depth of 3,028
feet. During the descents, Beebe and
Barton made detailed observations of
countless marine creatures and studied
the penetration of light in the sea.

BOND, GEORGE FOOTE
born 1915　　　　　　　United States
1954: Became diving and submarine
medical officer in the U.S. Navy.
1957–1963: Conducted a series of
experimental dives in a compression
chamber at the Naval Medical Research
Laboratory, Connecticut. Studied the
effect on divers of breathing oxygen-
helium mixtures for long periods at
depths in excess of 200 feet.
1964–1965: Headed U.S. Navy's
underwater habitat projects *Sealab I*
and *Sealab II*. The first of these
habitats was set 192 feet down, and
4 divers lived in it for 10 days. The
second operated at 206 feet and
involved three 10-man teams living in
the habitat for 15- and 30-day periods.
1966–1971: Continues as special
projects officer for undersea research.

BRUUN, ANTON FREDERIK
1901–1961　　　　　　　　Denmark
1928–1930: Sailed as a zoologist with
Johannes Schmidt on the last of a
series of Danish voyages aboard the
research ship *Dana*.
1950–1952: Led an expedition aboard
the *Galathea* to explore the ocean
trenches and find whether animals lived
in the deepest parts of the sea. Trawled
in the Pacific, the Caribbean, and the
Atlantic, discovering many marine
creatures never before seen by man.
The deepest catch came from 33,678
feet down in the Mindanao Trench,
northeast of Mindanao Island in the
Philippines.
1959–1961: Took part in a Scripps
Institution of Oceanography expedition
to the South China Sea.
See Galathea *map on page 137*

BUSHNELL, DAVID
1742(?)–1824　　　　　　United States
1771: Designed a one-man submarine
while a student at Yale University.
1776: Completed the construction of
his submarine, calling it the *Turtle*. This
submarine, used in the Revolutionary
War, made an unsuccessful attack on
the British warship H.M.S. *Eagle* in
New York harbor.
1783: Having failed to gain financial
backing for his submarine, went to
Georgia, where he worked as a doctor
for the rest of his life.

COUSTEAU, JACQUES-YVES
born 1910　　　　　　　　　France
1936–1939: Began skin diving in the
Mediterranean. Experimented with
oxygen breathing apparatus.
1941: Made the first documentary film
of underwater fishing.
1943: Tested the first Aqua-Lung,
which he had invented with Émile
Gagnan. In his first dive with the Aqua-
Lung reached 60 feet. With Frédéric
Dumas and Philippe Tailliez, made 500
successful dives to depths between
50 and 100 feet. Record dive of this
series was to 210 feet.
1945: Founded the French Navy's
Undersea Study and Research Group.
1947: Broke all previous depth records
with a dive to 297 feet.
1948: Served as security officer aboard
the oceanographic vessel *Élie Monnier*
during the first unmanned dive of
Auguste Piccard's bathyscaph.
1949–1961: Commanded the research
ship *Calypso* on numerous expeditions
to explore and film the undersea world.
Pioneered the development of new
devices to aid underwater research.
1962: Set up the world's first undersea
habitat, *Conshelf I*. Two divers lived
there for a week at a depth of 33 feet.
1963: Developed *Conshelf II*, a group
of underwater dwellings. One settlement, at 36 feet, housed 5 men for a

month. Another, at 90 feet, was occupied by 2 divers for a week.
1965: Headed the *Conshelf III* project involving 6 divers who spent 22 days at a depth of 330 feet.
1971: Continues to devote himself to undersea research. Is director of the Oceanographic Museum in Monaco, a post that he has held since 1957.
See entries under Dumas and Tailliez
See map on page 92

DUMAS, FRÉDÉRIC
1900's   France
1937–1939: Met Cousteau and Tailliez while catching fish in the Mediterranean. The three men continued diving as a team.
1943: Reached a record depth of 210 feet with the Aqua-Lung. Encountered for the first time the effects of nitrogen narcosis.
1945: Became a member of the French Navy's Undersea Study and Research Group. His early work with the group included important studies of the effects of explosions underwater.
1948: Planned diving courses for Aqua-Lung divers. Sailed aboard the Undersea Research Group's ship *Élie Monnier*. While investigating sunken ships, dived to a record 306 feet.
1949–1961: Sailed many times aboard the *Calypso*. Continues underwater research.
See entry under Cousteau
See map on page 92

EWING, MAURICE
born 1906   United States
1934: Began his oceanographic career by prospecting for oil in the Gulf of Mexico. As a result of this work, he became interested in ocean sediments and in the earth's crust beneath the sea floor. He used explosives to study the continental shelf from the deck of an open whale-boat.
1940–1944: Continued his studies aboard the Woods Hole Oceanographic Institution's vessel *Atlantis*.
1948: Set up the Lamont-Doherty Geological Observatory of Columbia University which he still directs.
1953: Purchased the schooner *Vema* and fitted her out as a research ship.

1954–1960: Sailed on many expeditions aboard the *Vema*, using seismic techniques to explore the ocean bottom and the layers beneath. Invented a piston corer for obtaining sediment samples. Discovered the extent of the mid-ocean ridge system.
1968: With co-chief scientist J. Lamar Worzel, sailed aboard the *Glomar Challenger* on her first drilling project. During the voyage, the *Glomar* drilled into the Sigsbee Knolls in the Gulf of Mexico, finding that they were composed of salt as Ewing and Worzel had previously maintained.
See Glomar *map on page 115*

FARGUES, MAURICE
(?)–1947   France
1945: After many years as a diving instructor and experimenter with hard hat equipment, he joined Cousteau in the Undersea Study and Research Group. Continued training divers with the group. Took part in numerous dives, and was in charge of the group's emergency supply launch.
1947: Died after reaching a record depth of 396 feet.
See map on page 92

FORBES, EDWARD
1815–1854   England
1841–1842: Sailed as a naturalist aboard the British survey ship *Beacon* on a voyage in the Mediterranean and Aegean seas. Dredged the bottom at 1,380 feet, deeper than any previous scientist. Classified animal and plant life of the sea into four zones according to depth. Believed that no life existed at depths beyond 1,800 feet.

FULTON, ROBERT
1765–1815   United States
1800: Built a submarine called the *Nautilus*. This vessel was driven by sail on the surface and by hand-operated propeller underwater. Carried out successful trials with the *Nautilus* and with another submarine called the *Mute* while living in France, but failed to interest the French government in his inventions.
1806: Returned to the United States where he built the first commercially

successful steamboat, *Clermont*.

GAGNAN, ÉMILE
1900's   France
See entry under Cousteau

HALLEY, EDMUND
1656–1742   England
1690: Developed a method of pumping pressurized air into a diving bell. Designed a diving bell using this technique, and tested it successfully in depths of around 60 feet. Devised a prototype diving suit by supplying divers with individual bells linked to the main bell by breathing tubes.

HARDY, SIR ALISTER
born 1896   England
1924: Sailed with Johan Hjort aboard the *Michael Sars* to study whales and plankton off Norway and Iceland.
1925–1927: Was chief zoologist on the research ship *Discovery* during its expedition to the South Atlantic. Carried out extensive research into whale behavior. Invented a continuous plankton recorder for obtaining plankton samples of which he made a detailed study. After the voyage, he continued his investigations of underwater life and has published a number of books on marine biology.
See Discovery *map on page 129*

HJORT, JOHAN
1869–1948   Norway
1892–1894: Began research in the Norwegian Department of Fisheries. Made a detailed study of sea squirts.
1899: At his suggestion, the Norwegian government built the vessel *Michael Sars* for fishery research.
1900–1909: Sailed aboard the *Michael Sars* on a series of scientific expeditions in the Norwegian Sea and the northern North Sea.
1910: Took the *Michael Sars* on an extended cruise to the Mediterranean, the Canary Islands, the Azores, and Newfoundland. Made wide biological and other oceanographic observations. Developed a new type of tow net that could be closed before being hauled to the surface.
1914–1915: Conducted fishery re-

search on behalf of the Canadian government.
1924–1939: President of the International Council for the Exploration of the Sea. Undertook further voyages, making particularly important studies of fish migration. Helped to prepare new laws for whale conservation.

HOLLAND, JOHN PHILIP
1840–1914   Ireland
1858–1872: Designed a submarine which was to be powered by a gasoline engine on the surface and by electricity underwater.
1873: Settled in the United States. Failed to interest the U.S. Navy in his submarine but was given financial support by a group of Irish patriots who hoped to use the submarine against the British Navy.
1878–1881: Built two submarines, the second of which made successful tests.
1888–1895: Commissioned by the U.S. Navy to build a new submarine but this vessel proved unsatisfactory.
1898: Built the submarine *Holland* with private funds. This craft became the U.S. Navy's first submarine, and Holland continued to build many more such vessels for the navy.

KEMP, SIR STANLEY
1882–1945   Ireland
1903–1909: As a naturalist in the Irish Department of Agriculture, he studied undersea life dredged from depths of 1,800 to 6,000 feet in the continental shelf area off west and southwest Ireland.
1910–1924: Went to India, where he studied marine and freshwater life, discovering new species of fish.
1925–1927: Led an expedition aboard the British research ship *Discovery* to study whales, plankton, and other marine life in the South Atlantic.
1929–1931: Sailed as leader on the first of a series of scientific voyages made by the *Discovery II*. During this expedition, extensive oceanographic research was carried out around the Antarctic continent.
1936–1945: Director of the Marine Laboratory at Plymouth, England.
See Discovery *map on page 129*

LETHBRIDGE, JOHN
(?)–1759   England
1715: Made his first dive in a barrel-like diving suit of his own invention.
1716–1749: Employed by London merchants to salvage treasure from sunken ships. Dived to numerous wrecks off the coasts of England, Holland, and Spain. He made over 100 dives to depths of around 60 feet. His deepest dive is reported to have been to 72 feet. He came near to drowning on at least five occasions.

LINK, EDWIN ALBERT
born 1904   United States
1929: Invented a mechanical trainer for aircraft pilots.
1962: Developed an aluminum cylinder which became the first underwater habitat to operate at considerable depth. It was sited 200 feet down in the Mediterranean, and marked the start of the *Man in Sea* project.
1964: Invented an undersea habitat known as SPID (Submersible Portable Inflatable Dwelling). The SPID was tested at 432 feet off the Bahamas by Robert Sténuit and Jon Lindbergh.
1965: Headed a further stage of the *Man in Sea* project during which two divers tested a pressure chamber for 48 hours at a depth of 650 feet. The experiment took place at the Linde Laboratories, Tonawanda, New York.
1967: Designed the *Deep Diver,* an undersea workboat incorporating a pressurized chamber and one at atmospheric pressure.
See entry under Sténuit

MAURY, MATTHEW FONTAINE
1806–1873   United States
1825–1839: Served in the U.S. Navy. Began to study winds and currents as a means to improve ocean travel.
1842–1852: Took charge of the Navy's Depot of Charts and Instruments. Studied data from hundreds of voyages and prepared charts indicating fastest and safest sea routes.
1853: Attended a maritime conference in Belgium. Helped to set up an international system of meteorological observation.
1855: Published *Physical Geography of*

the Sea and Its Meteorology, containing many theories about the basic processes of the sea.
1866: With John M. Brooke, developed a sounding device that could take a sample of the ocean floor. Organized soundings of the North Atlantic which enabled the first transatlantic telegraph cable to be laid.

MILNE-EDWARDS, HENRI
1800–1885   France
1844: Became the first scientist to explore the ocean floor in a diving suit. Using an early type of hard hat suit, he collected specimens of marine life off the coast of Sicily. Made many more dives in the Mediterranean, studying in particular crustaceans, mollusks, and marine worms.

NANSEN, FRIDTJOF
1861–1930   Norway
1882: Sailed as a zoologist on a voyage to the Arctic aboard a whaler.
1893–1896: Led an expedition to the Arctic in the *Fram*. When the vessel became locked in ice, Nansen and his crew made many valuable oceanographic, astronomical, and meteorological observations. They sounded depths of over 12,000 feet, measured the temperature and studied underwater life. Nansen left the ship and tried to sledge to the North Pole. He was forced back, and returned to Norway aboard a British ship.

PÉRON, FRANÇOIS
1775–1810   France
1800–1804: Sailed as a naturalist on a French voyage around the world. Measured the temperature of the sea at various depths. Finding that the water became increasingly cold with depth, he concluded that the ocean floor was covered with ice and that no life could exist there.

PHIPPS, CONSTANTINE JOHN
1744–1792   England
1773: While on a voyage toward the North Pole on board H.M.S. *Racehorse,* made the first deep sounding beyond the continental shelf in an ocean basin

between Iceland and Norway, where he
measured a depth of 4,098 feet.

PICCARD, AUGUSTE
1884–1962   Switzerland
1932: Ascended into the stratosphere
to a record height of 53,139 feet in a
balloon and gondola of his own
invention.
1933–1939: Designed the first bathyscaph for deep-sea exploration.
1945–1947: With financial backing
from the Belgian National Fund for
Scientific Research, completed the
construction of the bathyscaph *FNRS 2.*
1948: Accompanied by French naturalist Théodore Monod, made the first
shallow test dive in the bathyscaph to
84 feet. On the bathyscaph's first deep
dive, it descended, unmanned, to a
depth of 4,554 feet in the Atlantic
Ocean off Dakar, West Africa.
1950: Acted as adviser during the early
stages of the construction of a French
Navy bathyscaph, the *FNRS 3.*
1952: Went to Trieste, Italy, with his
son Jacques to build a new bathyscaph,
the *Trieste,* with the aid of Swiss and
Italian funds.
1953–1958: With Jacques Piccard,
made a number of successful dives in
the *Trieste.* The deepest descent was to
10,300 feet in the Tyrrhenian Sea, west
of Naples.
1958: *Trieste* purchased by the United
States Office of Naval Research.

PICCARD, JACQUES
born 1922   Switzerland
1948: Took part in the testing of the
first bathyscaph, designed by his father,
Auguste Piccard.
1952: Supervised the construction of
the bathyscaph *Trieste.*
1953–1958: With his father made a
number of dives in the *Trieste.*
1958: *Trieste* bought by the U.S. Office
of Naval Research. Piccard continued
his dives, now with U.S. scientists and
naval officers. On one dive reached
23,000 feet.
1960: With Donald Walsh of the U.S.
Navy, made a record-breaking descent
in the *Trieste* to 35,800 feet in the
Mariana Trench, southwest of Guam
in the Pacific.
1963–1964: Designed and constructed
the first mesoscaph, a middle-depth
underwater vehicle.
1969: Led a scientific team on the Gulf
Stream Drift Mission in the *Ben
Franklin,* a research mesoscaph of his
own design.
See Ben Franklin *map on page 80*

POSIDONIUS
135 B.C.(?)–51 B.C.(?)   Syria
97 B.C.: Taught Stoic philosophy in
Rhodes (an island in the Aegean Sea).
Traveled in Egypt, Spain, and France.

While sailing toward Spain in order to
reach the Atlantic, he made a depth
measurement of 6,000 feet near the
coast of Sardinia.

PYTHEAS
dates unknown   Greece
about 325 B.C.: Left Massalia (presen:-
day Marseille), sailed around the Iberian
Peninsula and on to Britain, which he
circumnavigated and explored. Continued northward until he reached the
frozen seas of the Arctic. Made many
scientific observations during the
voyage. Believed the ebb and flow of
tides to be related to the moon.

ROSS, SIR JAMES CLARK
1800–1862   Scotland
1839–1843: Led an expedition to the
Antarctic aboard H.M.S. *Erebus* and
H.M.S. *Terror.* Made the first extensive
series of deep-sea soundings, the
deepest of which was 14,500 feet.
Took deep-sea temperature measurements. Reached latitude 78° 10′ south,
the farthest south anyone had
penetrated before 1900.
See entry under Ross, Sir John

ROSS, SIR JOHN
1777–1856   Scotland
1818: Sailed from England, accompanied by his nephew James Clark
Ross, on a voyage of discovery to the
Arctic Ocean, seeking a Northwest
Passage to Asia. Made soundings and
took samples of the ocean floor. Found
marine life at a depth of 6,000 feet.
1829: Made a second unsuccessful
attempt to find a Northwest Passage.

SCHMIDT, JOHANNES
1877–1933   Denmark
1902–1904: Sailed aboard the research
ship *Thor* to study food fishes in the
North Atlantic. Found an eel larva in
deep water off the Faeroe Islands.
1905–1920: Went on a series of expeditions in the *Thor,* making an extensive
study of eel migrations.
1920–1922: During a voyage aboard
the research vessel *Dana,* he established
that eels migrate to a spawning ground
in the Sargasso Sea, a large region in
the North Atlantic Ocean.
1924–1930: Led a series of around-the-
world oceanographic expeditions in
the *Dana,* making detailed observations
of marine life.

SIEBE, AUGUSTUS
1788–1872   Germany
1816: Went to England where he
worked as an engineer.
1819: Invented the prototype of the
modern helmet diving suit.
1837: Developed his earlier invention

into a full diving suit, consisting of a
copper helmet screwed onto a rubber
suit, with lead-soled boots, and ballast
carried on the chest.

STÉNUIT, ROBERT
born 1933   Belgium
1962: Became chief diver of the *Man
in Sea* project. Tested Edwin Link's
first underwater habitat, in which he
spent 26 hours at a depth of 200 feet.
1964: With Jon Lindbergh (son of
American aviator Charles Lindbergh),
lived in Link's SPID habitat for 49 hours
at 432 feet. Continues test diving and
is a salvage expert.
See entry under Link

TAILLIEZ, PHILIPPE
born 1905   France
1937–1939: Joined with Cousteau and
Dumas on dives in the Mediterranean.
Experimented with underwater photo-
graphy and developed an undersea gun.
1943: Helped test the first Aqua-Lung.
1945: Became a founder member of the
Undersea Study and Research Group.
1948: Acted as communications diver
during the first shallow test dive of
Auguste Piccard's bathyscaph. With
Cousteau, sailed aboard the *Élie
Monnier* on the Undersea Research
Group's first major expedition.
1954: Commanded surface operations
for the descent of the French bathyscaph
FNRS 3 to a depth of 13,287 feet.
See entry under Cousteau
See map on page 92

THOMSON, SIR CHARLES
WYVILLE
1830–1882   Scotland
1868: Sailed on a dredging expedition
aboard H.M.S. *Lightning.* Studied
undersea life in the Atlantic from the
Faeroe Islands, north of Scotland, to
Gibraltar.
1869–1870: Went on a second voyage
in the Atlantic in H.M.S. *Porcupine.*
Dredged living creatures from depths
in excess of 15,000 feet.
1872–1876: Led a scientific team
aboard H.M.S. *Challenger* on the first
global oceanographic expedition.
During the voyage, which took the
Challenger through every ocean except
the Arctic, Thomson and his team made
extensive biological, chemical, geological, and physical observations. The
data they collected filled 50 volumes of
the *Challenger Reports.*
See Challenger *map on page 40*

WALSH, DONALD
born 1931   United States
See entry under Piccard, Jacques

WORZEL, J. LAMAR
born 1919   United States
See entry under Ewing

Glossary

abyss: Deep-sea region never penetrated by sunlight, extending from about 6,600 feet beneath the surface to the ocean floor. The abyss is the largest region in the sea and covers about half the earth's surface.

algae: Plant group to which most sea plants belong. Algae also grow in fresh water and in damp places on land. They are generally classified into five main groups: blue-green, green, brown, and red algae, and diatoms. The larger brown and red algae found in the sea are usually known as seaweeds. Marine algae may grow on stones and rocks or drift in the ocean. Some species live on other plants or cling to animals. *See also diatom*

atmosphere: A unit of pressure equal to 14·7 pounds per square inch (the average pressure of air at sea level). Pressure in the sea increases by one atmosphere for every 33 feet of depth.

baleen: Plates of whalebone growing downward like curtains from the upper jaw of certain whales, in place of teeth. The whale takes in gulps of water and squeezes them out through the baleen. Food is trapped in a fringe of bristles on the inner edge of the baleen and is then licked off. Among the baleen whales are the finback, the humpback, the sei, and the largest animal in the world – the huge blue whale.

bends: *See decompression*

Continental Drift: Theory that the continents are moving on the surface of the earth. According to this theory, the continents were once joined in one huge land mass. This land mass split into several parts, which then drifted apart. Scientists now think that the continents may be drifting apart as much as six inches every year.

continental shelf: The relatively shallow part of the seabed which borders most continents. The continental shelf slopes gradually from the shore to a maximum depth of about 600 feet. The shelf may extend for hundreds of miles where the coast is backed by lowlands, but in places where the coast is mountainous the shelf may be very narrow. The water above the continental shelf is rich in fish and the shelf contains valuable oil and mineral deposits.

core: A cylinder of sediment from the ocean floor. A core is obtained by driving a hollow tube as deeply as possible into the seabed, usually by means of a heavy lead weight. Modern coring devices use a movable piston to suck the sediments into the tube without disturbing their natural layering. Some of the longest cores taken from the sea floor measure more than 70 feet in length.

crustacean: A shell-covered animal with a segmented body, jointed legs, and feelers. Crustaceans belong to a large group of animals with jointed legs known as *arthropods*. There are about 30,000 kinds of crustaceans. Commonly known as shellfish, they include such creatures as lobsters, crabs, shrimps, and barnacles.

decompression: The reduction of pressure. When a diver returns to the surface he must decompress gradually by ascending slowly and making a number of stops according to a decompression table. If he ascends too fast he may suffer from *decompression sickness* or the *bends*. This condition is caused by rapid release of pressure, which makes the gas in blood and body tissues take the form of bubbles. These bubbles can block the blood flow and cause pain or, in some cases, permanent injury or even death. The bends may be treated by recompressing the diver in a pressure chamber, then lowering the pressure gradually.

decompression sickness: *See decompression*

deep scattering layer: An undersea phenomenon which appears as scattered markings on an echo-sounding device. Discovered in the early 1940's, this layer is thought to be a dense blanket of marine organisms hovering in the ocean. Its depth varies according to the migration of sea creatures which move toward the surface at night and descend at dawn. Several such layers, usually between 150 and 600 feet thick, may be found in depths ranging from 600 to 2,400 feet below the surface.

diatom: A microscopic water plant of the algae group. It consists of a single cell and has a shell largely composed of silica. The shells of diatoms form vast deposits on the sea floor and at the bottom of lakes. Golden-brown in color, diatoms may grow on stones, float freely, or move themselves through the water. They make up a large part of plankton and form a vital part of the sea's basic food supply. *See also plankton and algae*

dredge: An apparatus for obtaining specimens of underwater life and materials from the seabed. A simple dredge consists of a rigid frame to which a net is attached. The dredge is dragged across the sea bottom and traps materials that lie in its path. Some dredges consist of a series of buckets, or a single large grab, that dig into underwater deposits. Others suck up deposits by means of compressed air. More advanced dredges are being developed to recover minerals from the deep ocean floor.

ecosystem: Animal and plant populations, their relationship to one another, and to their environment. The life of a living organism is based on its ecosystem, which includes food supply, weather, and so on.

fathom: A unit of length used for measuring the depth of water. One fathom equals six feet, or 1·8 meters.

inner space: The space beneath the surface of the sea.

knot: A unit of speed used in sea and air navigation. It is equal to one nautical mile (1·1508 land miles) per

hour. The term *knot* originates from the early use of a knotted line to measure the distance traveled by a ship in a given length of time.
See also nautical mile

meter: A unit of length equal to 39·37 inches (3·281 feet).

mollusk: An animal with a soft body and no bones. Most mollusks have a shell either outside or inside their bodies. They live chiefly in the sea or in damp places on land. There are about 100,000 known kinds of living mollusks. They are arranged into six main groups. Mollusks include snails, oysters, mussels, clams, octopuses, squids, and cuttlefish.

nautical mile: A unit of length used to measure distances on, or in, the sea. The nautical mile is calculated by dividing the earth's circumference into 360 degrees and each degree into 60 minutes. One nautical mile equals one minute or ·0000463 of the earth's circumference. It has been agreed internationally that one nautical mile should measure 1·1508 land miles.

nitrogen narcosis: Condition affecting a diver's mental processes apparently caused by the presence of too much nitrogen in his blood and body tissues. A diver suffering from nitrogen narcosis gradually loses his ability to reason and may commit acts which endanger his life. Cousteau called this condition *the rapture of the depths.* The effects of nitrogen narcosis increase with depth. They may begin soon after 60 feet, and by about 300 feet the diver is no longer able to perform even routine tasks. To counteract this problem, a mixture of helium and oxygen is generally used for deep dives in place of nitrogen and oxygen. The amount of oxygen in the mixture is also carefully regulated because too high a proportion of this gas can lead to oxygen poisoning.

oceanography: The science of the ocean, and the life in it.

pelagic sediments: Deposits of fine-grained material found on the deep ocean floor. These sediments consist of the remains of animals and plants and of inorganic material.
See also terrigenous sediments

plankton: The mass of microscopic, drifting life found in bodies of water. It is made up of tiny plants, known as phytoplankton, and small animals called the zooplankton. The most numerous members of the phytoplankton are the diatoms. The phytoplankton is the basis of the food pyramid of the sea, because all marine animals feed on it, either directly or indirectly. Plants use energy from sunlight in order to grow and, for this reason, phytoplankton is found in the sunlit waters near the surface. The zooplankton, however, may be found at most depths. Usually the zooplankton animals rise toward the surface at night to feed on the phytoplankton, and return to deeper levels at dawn.

rapture of the depths: *See nitrogen narcosis*

seismic: Of, or caused by, an earthquake. The word seismic can also be used to refer to the effect of any artificial vibration of the earth.

sonar: A device that uses sound to determine the depth of water or to detect underwater objects. The term comes from the words **so**und **n**avigation **a**nd **r**anging. Sonar equipment works by sending a sound wave toward the ocean bottom. When the wave meets an obstruction, or strikes the ocean floor, it reflects back. As the speed of sound in water is known to be 4,700 feet a second, the time taken for the wave to return can be used to calculate depth.

telechiric: An unmanned underwater vehicle which operates by remote control from the surface.

terrigenous sediments: Deposits of material on the seabed. These deposits consist of particles of all sizes which

come mainly from nearby land. Terrigenous sediments cover most of the continental shelf area.
See also pelagic sediments

thermocline: A layer of water separating bodies of water at different temperatures. The thermocline lies between the warmer surface waters and the colder water beneath. It is caused because water expands when it is heated and becomes lighter. This lighter water does not mix with colder water but floats on top of it. The depth of the thermocline varies at different times of the year according to how far heat penetrates into it. When the surface waters cool, the thermocline breaks down.

tide: The alternate rise and fall of ocean waters mostly occurring about once every 12 hours. Tides are caused by the pull of the moon and sun on the earth. The moon's gravity pulls up the water directly beneath it, forming a bulge of water, or high tide. At the same time, the moon pulls the solid earth away from the water on the opposite side of the earth, producing another bulge, and high tide there too. As the earth rotates on its axis, one tidal bulge remains under the moon, and the other on the opposite side of the earth. As the earth turns once on its axis every 24 hours, every place on the earth's surface will have two high tides during that period. At the full moon and the new moon the pull of the sun is added to the pull of the moon to produce particularly high tides. These are called spring tides. In each month there are also two neap tides, which are less high than normal tides. Neap tides occur when the forces of sun and moon pull at right angles to each other.

trawl: A large net which is pulled along through the water, usually for large-scale fishing. There are many different kinds of trawl, equipped with a variety of devices to control the depth of the trawl and to keep the mouth of the net wide open.

Index

Picture Credits

Listed below are the sources of all the illustrations in this book. To identify the source of a particular illustration, first find the relevant page on the diagram opposite. The number in black in the appropriate position on that page refers to the credit as listed below.

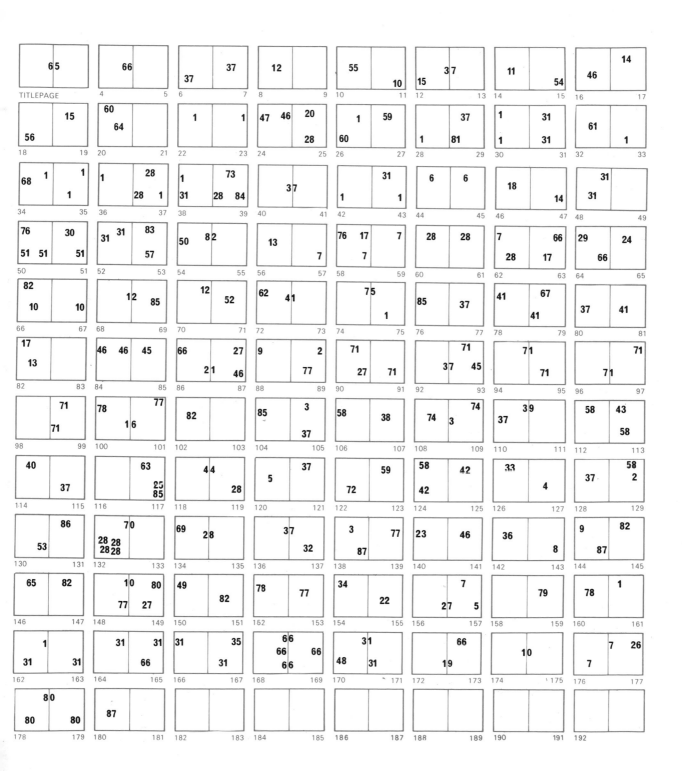